LABORATORY MANUAL

Introduction to Computer Science

Java Programming

by

Julie A. Anderson
Kathleen M. Austin
Lorraine N. Bergkvist

Publisher
The Goodheart-Willcox Company, Inc.
Tinley Park, Illinois
www.g-w.com

Copyright © 2021
by
The Goodheart-Willcox Company, Inc.

All rights reserved. No part of this work may be reproduced, stored, or transmitted in any form or by any electronic or mechanical means, including information storage and retrieval systems, without the prior written permission of The Goodheart-Willcox Company, Inc.

Manufactured in the United States of America.

ISBN 978-1-64564-181-0

1 2 3 4 5 6 7 8 9 – 21– 25 24 23 22 21 20

The Goodheart-Willcox Company, Inc. Brand Disclaimer: Brand names, company names, and illustrations for products and services included in this text are provided for educational purposes only and do not represent or imply endorsement or recommendation by the author or the publisher.

The Goodheart-Willcox Company, Inc. Safety Notice: The reader is expressly advised to carefully read, understand, and apply all safety precautions and warnings described in this book or that might also be indicated in undertaking the activities and exercises described herein to minimize risk of personal injury or injury to others. Common sense and good judgment should also be exercised and applied to help avoid all potential hazards. The reader should always refer to the appropriate manufacturer's technical information, directions, and recommendations; then proceed with care to follow specific equipment operating instructions. The reader should understand these notices and cautions are not exhaustive.

The publisher makes no warranty or representation whatsoever, either expressed or implied, including but not limited to equipment, procedures, and applications described or referred to herein, their quality, performance, merchantability, or fitness for a particular purpose. The publisher assumes no responsibility for any changes, errors, or omissions in this book. The publisher specifically disclaims any liability whatsoever, including any direct, indirect, incidental, consequential, special, or exemplary damages resulting, in whole or in part, from the reader's use or reliance upon the information, instructions, procedures, warnings, cautions, applications, or other matter contained in this book. The publisher assumes no responsibility for the activities of the reader.

The Goodheart-Willcox Company, Inc. Internet Disclaimer: The Internet resources and listings in this Goodheart-Willcox Publisher product are provided solely as a convenience to you. These resources and listings were reviewed at the time of publication to provide you with accurate, safe, and appropriate information. Goodheart-Willcox Publisher has no control over the referenced websites and, due to the dynamic nature of the Internet, is not responsible or liable for the content, products, or performance of links to other websites or resources. Goodheart-Willcox Publisher makes no representation, either expressed or implied, regarding the content of these websites, and such references do not constitute an endorsement or recommendation of the information or content presented. It is your responsibility to take all protective measures to guard against inappropriate content, viruses, or other destructive elements.

Cover image: Adrian Grosu/Shutterstock.com
Front matter design element: dynamic/Shutterstock.com

Contents

Introduction ix
    Using This Manual x
    General Safety Precautions x

Chapter 1 Computational Thinking 1
    Chapter Highlights 1
    Warm-Up Exercises 2
    Lab 1-1 Computational Thinking 3
        Exit Algorithm 4
        Grade-Level Vocabulary 5
        Dice Algorithm 6
        Debugging Challenge 7
    Lab 1-2 THINK 8
        Ancient Art of Tangrams 8
        Computerized Tangrams 10
        Debugging Challenge 11

Chapter 2 Encoding 13
    Chapter Highlights 13
    Warm-Up Exercises 14
    Lab 2-1 Hardware 15
        Four Components of Computers 16
        Debugging Challenge 17
    Lab 2-2 Software 17
        Operating Systems and Computers 17
        Debugging Challenge 19
    Lab 2-3 Binary Code 19
        Codes for Consistency 19
        Debugging Challenge 22

Chapter 3 Introduction to Java Programming 23
    Chapter Highlights 23
    Warm-Up Exercises 24
    Lab 3-1 Java Program Structure 25
        Exploring Unicode Characters 26
        Using Unicode Characters 27

    Using Non-Keyboard Characters 27
    Debugging Challenge 29
  Lab 3-2 Understanding Errors. .... .29
    Exploring jGRASP 30
    Debugging Challenge 32

Chapter 4 Variables 33
  Chapter Highlights 33
  Warm-Up Exercises 34
  Lab 4-1 Identifiers and Data Types 36
    Naming Variables.. 36
    Identifying Maximum and Minimum Values in Data Types 38
    Debugging Challenge 40
  Lab 4-2 Variable Values 40
    Writing Assignment Statements................................... 41
    Defining Constant Values.. 43
    Assigning Values from the User 44
    Debugging Challenge 46

Chapter 5 Java Expressions 47
  Chapter Highlights.................................47
  Warm-Up Exercises................................48
  Lab 5-1 Arithmetic Operators 51
    Calculating Numbers of Outfits 51
    Calculating a Grade Average 52
    Verifying Euler's Formula....................................... 54
    Debugging Challenge 56
  Lab 5-2 Operators and Expressions 56
    Calculating Constant Acceleration............................... 56
    Finding Force Due to Gravity.................................... 58
    How Fast Are You Running? 59
    Debugging Challenge 60
  Lab 5-3 Output Results 61
    Calculating Total Price 61
    Creating an Invoice 62
    Calculating Volume and Surface Area of a Sphere 63
    Debugging Challenge 64

Chapter 6 Classes 65
  Chapter Highlights 65
  Warm-Up Exercises 66
  Lab 6-1 Introduction to Classes 68
    Using a Game Object Class with a Client Class 68

- Exploring Turtle Graphics 70
- Making an Original Turtle Drawing 72
- Debugging Challenge 73
- **Lab 6-2 Java Class Library** 73
  - Using the Random Class 74
  - Using the Math Class 75
  - Debugging Challenge 77

Chapter 7 Drawing 79
- Chapter Highlights 79
- Warm-Up Exercises 80
- **Lab 7-1 Java Graphics Components** 82
  - Resizing the JavaFX Applications Window 82
  - Setting the Stage Style 83
  - Debugging Challenge 85
- **Lab 7-2 Text and Color** 85
  - Setting Properties for Text 86
  - Adjusting Color Brightness 87
  - Debugging Challenge 88
- **Lab 7-3 JavaFX Shapes** 89
  - Designing a Dominoes Game Piece 89
  - Making a Custom Button for an App 92
  - Debugging Challenge 94

Chapter 8 Selection 95
- Chapter Highlights 95
- Warm-Up Exercises 96
- **Lab 8-1 Conditions** 99
  - Using Comparisons to Determine Eligibility for a Driver's License 100
  - Using Comparisons to Determine If a Year Is a Leap Year 101
  - Debugging Challenge 103
- **Lab 8-2 Selection Statements** 104
  - Using Selection Statements to Rate Tornados 104
  - Using Selection Statements to Display Hours of Operation 106
  - Debugging Challenge 108
- **Lab 8-3 Helpful Conditions** 109
  - Using Nested Conditions for a Bike Rental App 109
  - Debugging Challenge 111
- **Lab 8-4 Comparing Objects** 111
  - Coding the All-Knowing Wizard 112
  - Comparing Values to Determine Output 113
  - Debugging Challenge 116

Chapter 9 Repetition..........................117
Chapter Highlights....................117
Warm-Up Exercises....................118
Lab 9-1 Java Loops....................121
Using a For Loop to Calculate Fibonacci Numbers..................121
Using a While Loop to Calculate the Total Price of a Market Basket...123
Using a Do/While Loop to Calculate a Square Root.................124
Debugging Challenge 126
Lab 9-2 Applying Loops...................126
Reading Data from a Bank Account Statement 126
Debugging Challenge 129

Chapter 10 String Processing 131
Chapter Highlights....................131
Warm-Up Exercises....................132
Lab 10-1 Creating Strings 134
Creating Strings to Add Line Numbers to a Printout 135
Debugging Challenge 136
Lab 10-2 String Methods...................137
Rearranging Strings..137
Debugging Challenge 142
Lab 10-3 Processing Strings Character by Character......142
Manipulating Strings to Convert Binary to Decimal 142
Debugging Challenge 146

Chapter 11 Managing Input and Output............147
Chapter Highlights....................147
Warm-Up Exercises....................148
Lab 11-1 Handling Exceptions 150
Validating Input for a Division-Practice Game 151
Reading from a File for a Number Game 152
Validating Input of Money...154
Debugging Challenge 155
Lab 11-2 Data Validation and Output..................156
Capturing and Validating a Birth Date 156
Reading from a File to Identify Near-Earth Asteroids 159
Debugging Challenge 162

Chapter 12 Custom Classes and Methods 163
Chapter Highlights....................163
Warm-Up Exercises....................164

| Lab 12-1 Creating Classes | 166 |

Creating a Player Class 166
Debugging Challenge 169

| Lab 12-2 Defining Methods | 170 |

Writing Methods to Complete a Class 170
Debugging Challenge 175

| Lab 12-3 Creating a Graphical Class | 176 |

Constructing a Ball Class 176
Debugging Challenge 182

Chapter 13 Working with Arrays 183

Chapter Highlights 183
Warm-Up Exercises 184
Lab 13-1 Creating Arrays 186

Creating a Sentence Analyzer Class Using an Array 186
Debugging Challenge 190

Lab 13-2 Processing Arrays 191

Using an Array for a Temperature Analyzer 191
Debugging Challenge 194

Lab 13-3 Searching Arrays 194

Using an Array to Find Popular Baby Names 194
Debugging Challenge 197

Chapter 14 Graphical User Interface 199

Chapter Highlights 199
Warm-Up Exercises 200
Lab 14-1 JavaFX Graphical User Interfaces 202

Designing an Age Calculator with UI Controls 202
Using Events to Animate Images 205
Animating Bouncing Objects 206
Debugging Challenge 207

Lab 14-2 GUI Input 208

Designing a Tip Calculator with UI Controls 208
Debugging Challenge 212

Lab 14-3 JavaFX Games 212

Adding a GUI to Create a Game App 213
Debugging Challenge 216

Chapter 15 Careers in Computer Programming 217

Chapter Highlights 217
Warm-Up Exercises 218

Lab 15-1 Benefits of Careers in Coding 219
Surveying Technology Careers 220
Problem-Solving with Small Groups 221
Problem-Solving with a Crowd 222

Lab 15-2 Preparation for Careers in Coding 224
Writing a Résumé Objective 224
Professional Accomplishments and Work History 225
References 227

Chapter 16 Computing and Society 229

Chapter Highlights 229
Warm-Up Exercises 230
Lab 16-1 Computing and Ethics 231
Ten Commandments of Computer Ethics 232
Privacy 233

Lab 16-2 Computing and Security 234
Protecting Your Data 234

Lab 16-3 Safe Computing 236
Privacy Policy 236
Evaluating a Website 237
Password Strength 237

Introduction

    This laboratory manual complements the *Introduction to Computer Science: Java Programming* textbook and classroom-related studies. The laboratory activities in this manual help develop the valuable skills needed to pursue a career in the computer science field as a Java programmer. Laboratory activities should be an essential part of your training. They link the concepts presented in the textbook to hands-on performance. You should not expect to learn Java programming skills only through the textbook, lectures, and demonstrations.

    Java is an easy-to-learn programming language. By using this laboratory manual along with studying the *Introduction to Computer Science: Java Programming* textbook, you will explore, you will experiment, and you will learn. This will open the window to the possibilities that lie ahead of you in the field of computer programming.

Using This Manual

The activities in the *Introduction to Computer Science: Java Programming* laboratory manual correlate to the textbook chapters. Each chapter has at least two laboratory activities, each of which begin with a brief overview of the activity. In some cases, this overview also sets up a scenario that will be used for the activity. Following the overview are learning goals. These are the objectives to meet by completing the activity.

After the learning goals, there is a list of the materials needed for the activity. Some activities can be completed with basic Internet access and a word processor. Other activities are more involved and require use of the jGRASP Integrated Development Environment and starter files downloaded from the student companion website. The list of materials will indicate what is needed so you can be prepared prior to starting the activity.

Following the list of materials are the step-by-step instructions for completing the activity. Read through the entire activity before starting to work through it. Ask your instructor for help if you have any questions or need clarification on any of the steps.

General Safety Procedures

- Do not attempt to make changes or modifications on machines you do not own or have permission to modify.
- Do not install unauthorized software on computers. Abide by all software-licensing agreements.
- Do not attempt to or access systems beyond those you have permission to access. Doing so may violate school policies or local or federal laws.
- Ensure your work does not needlessly interfere with the work of other students.

Name _____ Date _____ Class _____

CHAPTER 1
Computational Thinking

As the programmer, you control the behavior of a computer and make it do anything you say using its built-in instructions. Organizing these built-in instructions into computer applications is a human endeavor called computer programming. Thinking like a computer is called computational thinking. It takes computational thinking to translate human ideas into computer instructions.

Chapter Highlights

- The ways in which humans think and computers operate are vastly different.
- The components of computational thinking are decomposition, pattern recognition, pattern abstraction, and algorithm development.
- Software development is the process of identifying a need, specifying the components for a solution, formulating a design, writing the programs, developing documentation, writing test plans, executing test plans, and revising the software as a result of the testing.
- Successful programmers are confident they can find a solution and are good computational thinkers and relentless checkers.

While studying, look for the activity icon for:
- Vocabulary terms with e-flash cards and matching activities.
- Starter files for lab activities.

These activities can be accessed at
www.g-wlearning.com/informationtechnology/1773

Warm-Up Exercises

_____ 1. How are problem-solving and computational thinking related?
   A. Computational thinking and problem-solving are identical.
   B. Computational thinking is a subset of problem-solving.
   C. Problem-solving is a subset of computational thinking.
   D. Computational thinking and problem-solving have nothing in common.
   E. Both computational thinking and problem-solving are the ways in which computers operate.

_____ 2. Who is generally considered to be the first computer programmer?
   A. Ada Lady Lovelace
   B. Grace Hopper
   C. Katherine Johnson
   D. Herman Hollerith
   E. Bill Gates

_____ 3. Which of the following is not a trait of a successful programmer?
   A. positive attitude
   B. great communicator
   C. good time manager
   D. carelessness
   E. relentless checker

_____ 4. Choose the type of processing that can only be done by humans.
   A. financial calculation
   B. original thought
   C. storing information
   D. video playback
   E. encoding text data

_____ 5. Which component of computational thinking is about finding repeating schemes in problem-solving?
   A. decomposition
   B. pattern recognition
   C. algorithm
   D. abstraction
   E. computational thinking

_____ 6. The component of computational thinking that consists of breaking apart a problem into small parts is _____.
   A. decomposition
   B. pattern recognition
   C. algorithm
   D. abstraction
   E. computational thinking

_____ 7. Which component of computational thinking is about generalization of patterns (algebra)?
  A. decomposition
  B. pattern recognition
  C. algorithm
  D. abstraction
  E. computational thinking

_____ 8. The component of computational thinking that consists of writing instructions for a computer is _____.
  A. decomposition
  B. pattern recognition
  C. algorithm
  D. abstraction
  E. computational thinking

_____ 9. Which of the following problems *cannot* be solved by a computer?
  A. adding a table of numbers
  B. facial recognition
  C. medical history storage and retrieval
  D. keeping track of banking transactions
  E. world hunger

_____ 10. The most efficient algorithm is often referred to as _____.
  A. cumbersome
  B. abstract
  C. the most elegant
  D. pseudocode
  E. small, doable steps

Lab 1-1

Computational Thinking

Writing an algorithm is the last activity in computational thinking. It involves organizing the abstractions into the proper sequence to solve the problem. In this lab, you will practice computational thinking by examining and developing algorithms.

Learning Goals
- Apply the components of computational thinking.
- Compare and contrast human thought with computer activity.
- Write algorithms.

Materials
- Dice nets (instructor-supplied handout)
- Scissors
- Ballpoint pen
- Ruler or straightedge
- Glue or tape

Application and Extension of Knowledge

Exit Algorithm

Sooner or later, you will go home. There may be several ways to get there. This activity explores them in the format of an algorithm.

Consider you are in a classroom. Class has ended and it is time to leave. Write two different algorithms for getting around your school. Think like a computer, or perhaps a robot. For example, clearly state the steps to exit:

1. Collect your belongings.
2. Stand up.
3. Walk to the door.
4. Open the door if it is closed.
5. Walk into the hall.

Procedure

1. Write an algorithm to get from your classroom to the cafeteria.

2. Write an algorithm to get from the school to your home.

Reflections

1. Explain why you think these algorithms represent the best directions.

2. Obviously, you know that if a door is closed you need to open it to pass through. Explain why the algorithm in the example includes "open the door if it is closed."

Grade-Level Vocabulary

A computer can help you practice and learn new vocabulary. Great Schools, which is a nonprofit organization for parents of students, publishes lists of grade-level-appropriate vocabulary. Apply the steps of computational thinking to write an algorithm for learning these words.

Procedure

1. Launch a web browser, navigate to a search engine, and search for academic vocabulary for 10th graders. Write down the URL you found. Then, click through to the Great Schools website to view the list of words.

2. Apply step 1 of problem-solving: understand the problem. Your ultimate goal is to know all the words. You will develop an algorithm to solve this problem. In your own words, state the problem.

3. Apply step 1 of computational thinking: decompose. Break down the problem into small, doable steps. Consider that you may already know some of these words. List the steps you identify.

4. Apply step 2 of computational thinking: recognize patterns. Are there words you recognize, but do not know their meanings? Are there words you have never seen before? Describe the patterns you find.

5. Apply step 3 of computational thinking: generalize patterns. What might a computer be able to do for you over and over?

6. Apply step 4 of computational thinking: write an algorithm. Write an algorithm to practice the new words. Describe the steps a computer would follow to help you master this list.

Reflections

1. Explain if you would need a computer to help you master this list.

   _____
   _____
   _____

2. How would you use this knowledge of new words?

   _____
   _____
   _____

Dice Algorithm

In this activity, you will follow an algorithm to assemble two dice from the dice nets handout. Then, you and a partner will roll the dice and keep a tally of the numbers rolled.

Procedure

1. Follow the steps in this algorithm:
   A. Use a ballpoint pen and ruler to draw on the dotted line of each net in the dice nets handout. Press down so the pen makes an indentation. The goal is to crease the paper to allow for folding.
   B. Using scissors, cut on the solid lines around the net.
   C. Fold on the dotted lines so that the pips (dots) of the die are on the outside.
   D. Fold the net into a cube. Put glue on the tabs and glue them to the underside of the neighboring face of the die. If you are using tape, put the tabs under the faces of the die and put a small piece of tape on the edge.
   E. Roll the dice.
2. With a partner, roll one pair of dice 50 times. Make a tally in the table below next to the numbers that were rolled, and total the tally.

Roll	Tally Number of Times Rolled	Total of Tally
2		
3		
4		
5		
6		
7		
8		
9		
10		
11		
12		

Name _____   Chapter 1 Computational Thinking   7

Reflections

1. Comment on the ease of following the algorithm. Explain how you would make the directions better.

   _____
   _____
   _____

2. What is the most likely outcome when two dice are rolled? Hint: add two opposite faces of a die. How many ways can you roll a seven? What is the least-likely outcome? How many ways can you roll a two? Is there any other number that can be rolled only one way?

   _____
   _____
   _____

3. Explain if the way you roll the dice has a predicted outcome. For example, can you roll the dice to ensure the roll is a six and a three? Is there an algorithm for rolling specific numbers?

   _____
   _____
   _____

Debugging Challenge

Lorene gave directions to her friend Kate to find her house. This is the direction algorithm Lorene gave:

1. Start in the middle of town at Main Street and First Avenue.
2. Face the bank. That is north.
3. Go four streets east on First Avenue, and turn right. That is south on Bright House Lane.
4. Go two blocks to Berryhill Circle.
5. Go two thirds of the way around Berryhill Circle, and that is my house.

Kate got lost and texted Lorene for help. Why did Kate get lost? What would improve this algorithm? Record your response in the space provided. Draw a picture of the algorithm on a piece of scrap paper if you think that may help.

_____
_____
_____
_____
_____
_____
_____

Lab 1-2

THINK

Computers cannot think. Humans must do the thinking and then tell the computer what to do. In this lab, you will extend your knowledge of what a computer can do.

Learning Goals
- Describe how an algorithm is critical to a computer solution.
- Practice traits of successful programmers.

Materials
- Set of seven tans (instructor-supplied handout)
- Scissors

Application and Extension of Knowledge

Ancient Art of Tangrams

Tangrams are an ancient Chinese art form. Based on seven shapes, called tans, people rearrange the tans to make images of everyday objects. The solution is called a tangram. The following image shows a house being made from a set of wooden tans. The final tan needs to be moved into place to complete the tangram. Follow the directions below to write an algorithm to make a tangram.

Bankrx/Shutterstock.com

Procedure

1. Cut out the tans from the handout.
2. Lay out the tans on your workspace. Study them. Characterize the individual tans. How would you describe them?

_____

_____

_____

_____

Name _____

3. On the handout, the tans are positioned to form a square. Assemble your cut-out tans into a square following that configuration. Comment on the process below.

   _____
   _____
   _____
   _____
   _____

4. Write an algorithm to assemble the tans into a square. When you are finished, give the algorithm to a classmate to follow.

   _____
   _____
   _____
   _____
   _____
   _____
   _____
   _____
   _____
   _____
   _____

Reflections

1. Together with your classmate, discuss what happened when your algorithm was followed. Note anything that was unclear in your algorithm.

   _____
   _____
   _____
   _____
   _____

2. Speculate what problems might arise if you tried to write an algorithm for the computer to perform the same task.

   _____
   _____
   _____
   _____
   _____

Computerized Tangrams

In this activity, you will write an algorithm for a computer to arrange tans into an arrow in negative space. *Negative space* is an area that does not contain any objects, but is defined by objects. Notice how the arrow is formed in the following image. After writing the algorithm, you will see if a classmate can follow it like a computer.

Bankrx/Shutterstock.com

Procedure

1. Write the algorithm for a computer to place these tans to make the tangram of an arrow in a square.

2. Ask a classmate to test the algorithm for you. Describe the results.

Reflections

1. Write about the usefulness of the algorithm. What parts needed more information? Were there any parts that were too wordy? Were there any parts that were confusing?

   _____
   _____
   _____

2. Explain why the following image may be too difficult to program a computer to solve. What extra instructions would you need to solve this exact tangram by computer?

Bankrx/Shutterstock.com

   _____
   _____
   _____

3. Explain how you exhibited the traits of successful programmers in these exercises.

   _____
   _____
   _____

Debugging Challenge

There is an error in this algorithm to solve the cat tangram. What is wrong? Record your response in the space provided.

Making a Cat Tangram

1. Lay out all seven tans in front of you.
2. Select the two large right triangles.
3. Place one so the right angle is in the upper-right corner of the tan and the hypotenuse is going from lower left to upper right.

4. Place the second large triangle so the right angle is in the lower-right corner of the tan and place it next to the first tan at the top so the legs are touching.
5. Select the mid-sized triangle, and place it touching the first tan with the hypotenuse horizontal and the leg along the hypotenuse of the first tan meeting at the bottom vertex.
6. Select the parallelogram, and place it for the head.
7. Select the square tan, and place it so that one corner is touching the top vertex of the first tan and the square looks like a diamond.
8. Select the two small triangles, and place them alongside each of the square tan's lower two sides with the right angles meeting at the top of the square.

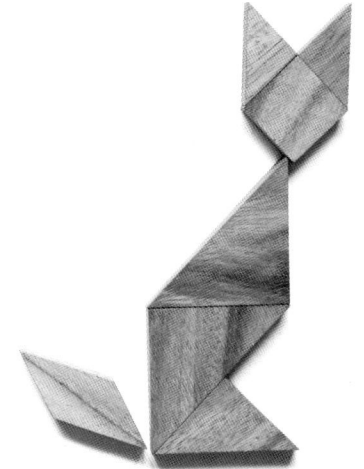

Bankrx/Shutterstock.com

Name _____ Date _____ Class _____

CHAPTER 2

Encoding

The physical parts of a computer are called hardware. Computer systems require specific directions to function. These directions or instructions, which are encoded, are known as software. Knowledge of number systems is essential to understanding how computers encode data.

Chapter Highlights

- The four major components of a computer system are: input, memory, processing, and output.
- It is the job of software to take the basic CPU instruction set and produce outcomes.
- Low-level programming languages assemble instructions into machine language.
- High-level programming languages use human-like language to create programs. Compilers or interpreters are then needed to translate the programs into machine language.
- Object-oriented programming creates code segments to keep data and instructions separated from the rest of the code.
- All digital computers use the binary number system to operate. Hexadecimal is a shorthand for writing binary numbers.
- Binary, decimal, and hexadecimal numbers are all positional number systems.
- All human-readable characters and non-readable codes are part of Unicode and ASCII.

While studying, look for the activity icon 📲 for:
- Vocabulary terms with e-flash cards and matching activities.
- Starter files for lab activities.

These activities can be accessed at
www.g-wlearning.com/informationtechnology/1773

Warm-Up Exercises

_____ 1. Two types of primary memory are _____.
   A. principal and secondary
   B. RAM and ROM
   C. CPU and LCD
   D. RAM and CPU
   E. ROM and storage

_____ 2. Using a printer for a document is considered _____.
   A. input
   B. processing
   C. storage
   D. output
   E. LCD

_____ 3. Magnetic, optical, and solid state are forms of _____.
   A. primary and secondary memory
   B. storage
   C. microprocessors
   D. chips
   E. input

_____ 4. Some programming languages have a limited set of recognizable English words. Which of these is *not* a high-level language?
   A. Java
   B. Visual Basic
   C. C++
   D. machine language
   E. COBOL

_____ 5. Which process is used to convert Java to machine language?
   A. executing binary code
   B. compiling source code
   C. interpreting bytecodes
   D. instantiation
   E. procedural language

_____ 6. Which is *not* true about the Unicode encoding system?
   A. It has advantages over ASCII.
   B. It has codes for emojis.
   C. It has codes for number systems.
   D. It is based on the English alphabet only.
   E. It is used by Java.

_____ 7. If the binary number 0011 were converted into a decimal number, it would be _____.
   A. 3
   B. 11
   C. 110
   D. 6
   E. 30
_____ 8. If the binary number 1110 were converted into a hex number, it would be _____.
   A. 14
   B. 11 10
   C. 16
   D. E
   E. A
_____ 9. If the binary number 1010 were converted into a hex number, it would be _____.
   A. 1010
   B. 101
   C. 16
   D. 10
   E. A
_____ 10. Select the memory type that cannot be changed.
   A. random-access-memory (RAM)
   B. read-only-memory (ROM)
   C. rewritable compact discs (CD)
   D. hard drive (HD)
   E. universal serial bus drives (USB)

Lab 2-1

Hardware

Four components are necessary for hardware to be categorized as a computer: a device that allows the user to input data in programs, a means to output data, a mechanism to process data, and a mechanism to store information. If one of these components is missing, the device is not considered a computer system. In this lab, you will apply what you learned about hardware to find information about your computer.

Learning Goals
- Identify the components of the computer.
- Identify input, memory, processing, and output devices on the computer.

Materials
- Computer

Application and Extension of Knowledge

Four Components of Computers

Computers appear to be a commodity. They all appear to be the same. You sit down in front of one and assume you can operate it. There are similarities among all personal computers. In actuality, computers are very different inside. This lab will explore the features of the components of your computer.

Procedure

1. Identify the make and model of your computer. Often this information is on the box that contains the CPU. If you are working with a portable computer, the information could be on the underside of the computer. Use the space provided to record the make and model of your computer.

   _____
   _____
   _____

2. If using a Windows computer, click in the search box on the task bar, enter control panel in the search text box, and select **Control Panel** in the list of results. In the **Control Panel** window, click **System and Security**, and then **System**. If using a Mac OS computer, click the Apple icon, and select About this Mac. For both operating systems, details about the processor, amount of installed memory, and system type are displayed. Record the results in the space provided.

   _____
   _____
   _____

3. A component of computer systems is input. In the space below, list the input devices available on your computer. Look for a keyboard, touch screen, mouse, stylus, microphone, and the like.

   _____
   _____
   _____

4. A component of computer systems is output. In the space below, list the output devices available on your computer. Look for a display, speakers, printers, and the like.

   _____
   _____
   _____

Reflections

1. Compare your findings in the previous activity with that of a classmate. Are your findings different from your classmates' findings? Write a statement to explain why this is or is not the case.

   _____
   _____
   _____

2. If you were unable to access the information, explain why not.

Debugging Challenge

Manuel wanted to save his homework file before submitting it. A CD-ROM was included in his textbook purchase. He thought it was a good place to save his work. That way, information from the course would all be in one place. He placed the disc in the CD drive and tried unsuccessfully to save his work to the CD-ROM. Why was he not able to do that?

Lab 2-2

Software

Software makes the hardware work. It contains all the instructions to operate the hardware. In this lab, you will apply what you learned about software to find information about your computer.

Learning Goals
- Identify the software components of the computer.
- Identify operating system information on the computer.

Materials
- Computer

Application and Extension of Knowledge

Operating Systems and Computers

Computers have operating systems unique to their instruction sets. Each version of a CPU has its own instruction set. While most of the instructions for a series of chips from a specific manufacturer may be similar, each new chip adds or deletes machine-level codes from the last iteration of the chip.

Procedure

1. Identify the CPU of your computer. Often this information is in the Control Panel. Use the space below to record the CPU chip name and number. Also record the processing speed, if available.

2. Some operating systems use 32 bits per instruction and others use 64 bits. In the same location as the CPU information is the bits per instruction (bpi). Record the bpi for your computer below.

_____
_____
_____

3. In the same location should be the type of operating system used in your computer. Locate that information, and record it in the space below.

_____
_____
_____

4. Identify four software programs installed on the computer. On Windows computers, this list is available in the **Start** menu. On Mac computers, look in the **Applications** folder.

_____
_____
_____

Reflections

1. Are your findings different from your classmates' findings? Write a statement to explain why this is or is not the case.

_____
_____
_____

2. Explain why different operating systems require different versions of software.

_____
_____
_____

Debugging Challenge

Yuan prepared a speech for his English class. He decided to talk about software. Help him by correcting the errors he has made in the speech. Read the following excerpt from his talk and record your response in the space provided.

Beginning programmers use machine languages until they are promoted to a higher position in the company. Then they use high-level languages. Examples of high-level languages are Bytecode, Python, C++, and C#. Often these languages are based on object-oriented programming. The CPU uses instantiation to handle instructions.

_____

_____

_____

_____

_____

Lab 2-3
Binary Code

Number systems used in computing include binary, octal, and hexadecimal. These have bases of two, eight, and sixteen digits. In this lab, you will apply your knowledge of number systems.

Learning Goals
- Encode text characters using Unicode.
- Convert numbers between decimal, binary, and hexadecimal.

Materials
- Computer
- Unicode characters chart (textbook Figure 2-9)

Application and Extension of Knowledge

Codes for Consistency

Binary and hex numbers are used in computers to encode text and other characters. In order to make computers usable in all parts of the world with different languages and cultures, a common universal code is necessary. Unicode specifies which binary and hex number patterns represent printable and nonprintable characters. It is the universal code. This activity applies what you learned about binary and hex to Unicode.

Procedure

1. Use the Unicode chart to decode and encode the following.

   A. Decode this Unicode decimal message into characters.
      0071 0111 0111 0100 0032 0076 0117 0099 0107 0033
      _____

   B. Decode this Unicode hex message into characters.
      004E 0041 0053 0041 002E 0047 004F 0056
      _____

   C. Encode this phrase in Unicode hex.
      You Win!
      _____

2. Use hand calculations or the computer's calculator to convert the following from decimal to hexadecimal.
   A. 32 _____
   B. 45 _____
   C. 50 _____
   D. 57 _____
   E. 62 _____
   F. 75 _____
   G. 84 _____
   H. 93 _____
   I. 106 _____

3. Use hand calculations or the computer's calculator to convert the following from decimal to binary.
   A. 37 _____
   B. 81 _____
   C. 90 _____
   D. 99 _____
   E. 103 _____
   F. 112 _____
   G. 120 _____
   H. 123 _____
   I. 126 _____

4. Use hand calculations or the computer's calculator to convert the following from binary to decimal.
   A. 0000 0000 0111   1001 0000 0000 0111

   B. 0000 0000 0111 1110

   C. 0000 0000 0010 1010

   D. 0000 0000 0100 1100   0000 0000 0100 1111   0000 0000 0100 1100

   E. 0000 0000 0011 1111

5. Use hand calculations or the computer's calculator to convert the following from binary to hexadecimal.
   A. 0000 0000 0010 0001   0000 0000 0111 0100

   B. 0000 0000 0010 1111   0000 0000 0111 1110

   C. 0000 0000 0101 0101   0000 0000 0010 1010

   D. 0000 0000 0101 1000   0000 0000 0110 1101

   E. 0000 0000 1000 1110   0000 0000 1100 0011

Reflections

1. Binary numbers are used in computers because the two digits can represent on and off for computers. Some people think that humans developed a base ten number system because we had ten fingers or digits. But, why not binary? After all, humans have two arms, two eyes, two hands, and so on. List advantages and disadvantages of the binary and decimal systems.

2. Explain why you think computer-software manufacturers would form a group to develop Unicode. After all, it is just a chart of codes. Any one person could do that.

Debugging Challenge

Ayesha decided to teach her little brother Amir how to count in binary. She had him write down the first ten Arabic numerals and told him these were from their heritage. She said that Arabs had devised these numerals. Then, she had him write the binary numbers below them. Review Amir's chart that follows. Then, help Ayesha find the errors in Amir's chart and record your response in the space provided.

```
0   1   2   3    4    5    6    7    8     9
0   1   11  10   100  101  110  111  1000  1010
```

_____

_____

_____

_____

_____

Name _____ Date _____ Class _____

CHAPTER 3
Introduction to Java Programming

This beginning look into the Java language—a simple program—using the jGRASP integrated development environment is a great way to learn basic information about programming in general and Java in particular. This lab applies what you have learned about writing a simple Java program and the information about Unicode. It also provides a peek into the compiler and its messages.

Chapter Highlights

- The Java language is property of Oracle Corporation.
- OpenJDK is a freely available, open-source version released under the GNU General Public License.
- The programmer writes Java source code in a plain-text file that is saved with a .java file extension. This is then fed into the Java compiler, which converts the source code into binary bytecodes. The compiler stores the bytecodes in a file with a .class file extension.
- To run the program, the Java Virtual Machine (JVM) interprets the bytecodes in the .class file. As the program executes, the JVM translates the bytecodes into the machine language of the computer on which the program is running.
- The *Java Development Kit (JDK)* is a set of tools for creating programs in Java. It includes the Java compiler, Java Virtual Machine, and lots of code that can be used to make your programs fun.
- An *integrated development environment (IDE)* is a program that makes it easy to write, compile, and run programs. This course uses jGRASP as the IDE.
- A *class* is a unit of code that works together. Each program is written as one or more classes.
- The main method indicates where the code for the program starts.
- *Blocks* are subsections of code.
- *Statements* perform the work of the program, such as performing a calculation or outputting information.
- Four types of errors that can occur when programming are compiler errors, runtime errors, logic errors, and user errors.

While studying, look for the activity icon for:
- Vocabulary terms with e-flash cards and matching activities.
- Starter files for lab activities.

These activities can be accessed at
www.g-wlearning.com/informationtechnology/1773

Warm-Up Exercises

_____ 1. The acronym JVM expands to:
   A. Java Victory Medal
   B. Joint Virtual Model
   C. Java Virtual Machine
   D. Junior Vault Method
   E. Java Valuable Machine

_____ 2. Java source code is written in a plain-text file using the _____ file extension.
   A. .jva
   B. .jvm
   C. .jdk
   D. .java
   E. .txt

_____ 3. An integrated development environment is a program that makes it easy to do all but which of the following?
   A. write
   B. compile
   C. edit
   D. run
   E. sell

_____ 4. The function of the main method of a Java program is to _____.
   A. be the starting point for the code for the program
   B. separate the class from the block
   C. keep the class public
   D. hold the block comments
   E. pass the comments to the compiler

_____ 5. In Java, the file name for the Java source code must match the _____.
   A. statement
   B. class name
   C. algorithm
   D. main method
   E. compiler message

_____ 6. Select the only error that is *not* a type of programming error.
   A. logic error
   B. compiler error
   C. timing error
   D. syntax error
   E. runtime error

_____ 7. Select the file extension the compiler uses to store the bytecodes.
   A. .java
   B. .exe
   C. .compiled
   D. .class
   E. .bytes

_____ 8. A logic error occurs when a programmer makes which type of mistake?
   A. typo in a keyword
   B. missing punctuation
   C. incorrect algorithm
   D. incorrect input
   E. removes static from the main method

_____ 9. The programmer is not responsible for user errors. However, a good programmer can take some of the following steps to minimize user errors. Which option does *not* belong in the list of actions?
   A. perform unit testing with a user
   B. notify the user of the error
   C. validate the input
   D. ignore user mistakes and keep processing
   E. ask again for correct input

_____ 10. A white space character is used to separate code. Which of the following is *not* a white space character?
   A. space
   B. tab
   C. new line character
   D. semicolon
   E. [Enter] key press

Lab 3-1

Java Program Structure

In the textbook, you created a simple Hello World! program. In this lab, you will apply what you have learned about writing a program in Java and using the Unicode characters.

Learning Goals
- Compile and execute a Java program.
- Create a simple Java program using Unicode.

Materials
- jGRASP Integrated Development Environment
- Starter files from the student companion website
- Web browser and Internet access

Application and Extension of Knowledge

Exploring Unicode Characters

In Chapter 2, you learned that Unicode is the standard for character representation in Java. This activity provides an opportunity to identify Unicode characters and their codes. You will then write your name using Unicode characters.

Procedure

1. Launch a web browser, and navigate to the Unicode website (www.unicode.org). Click on the **Code Charts** link at the bottom of the page under **Quick Links** to navigate to the Unicode Character Code Charts page. Locate the European Scripts section of the page, scroll down to the Latin heading, and click the **Basic Latin (ASCII)** link to open a PDF. Write down the URL and the title of the chart. What is the range of codes listed in this chart?

   _____
   _____
   _____

2. View the second page of the chart. You will see the ASCII characters and some control characters. Notice code 0007 in column 1. It is for a bell. Early computers were replacing typewriters. On a typewriter, when the carriage reached the end of a line and needed to be returned, a bell rang. On the early computers, pressing [Ctrl][G] rang the bell. Programmers could also ring the bell by printing the BEL character.

   Spend some time seeing how the rows and columns are organized on the first page of the chart. For example, look across line 7 and see that all character codes end in 7. The numbers across the top of the columns are the first three hex digits for the Unicode for a character and the row containing the character is the fourth digit. For example, consider the character G. The column code is 004. The row code is 7. Therefore, the Unicode for G is 0047. In the spaces below, write the column codes, the row codes, and the Unicode for these four characters: X, {, m, &.

   _____
   _____
   _____

Reflections

1. Look at the corresponding codes for uppercase and lowercase letters. Write an observation about the patterns in their codes.

   _____
   _____
   _____

2. Examine the digits. Write an observation about the pattern in their codes.

   _____
   _____
   _____

Name _____   Chapter 3 Introduction to Java Programming   27

Using Unicode Characters

In this activity, you will identify the Unicode characters for your name. Then, you will use these codes in a Java file to print your name.

Procedure

1. Examine the Unicode chart used in the previous activity. Notice that the uppercase letters and lowercase letters have different codes. Write the codes for your first and last names, using uppercase for the initial characters. Note that a space is Unicode 0020.

   _____

   _____

   _____

2. To use Unicode in Java, precede the code with \u. For example, the name Lakesia is encoded as

   ```
   "\u004C\u0061\u006B\u0065\u0073\u0069\u0061"
   ```

   Using Unicode in the print statement is performed as below.

   ```
   System.out.println( "\u004C\u0061\u006B\u0065\u0073\u0069\u0061" );
   ```

   In the space below, write the Java code to print the Unicode characters in your first and last names using \u. Do not forget to include the space between your names.

   _____

   _____

   _____

3. Launch jGRASP, and open the UnicodeName.java file downloaded from the student companion website.
4. Edit the System.out.println(" "); statement to insert your name between the quotation marks using Unicode.
5. Compile and run the program.

Reflections

1. Explain why this is *not* the best method for outputting letters.

   _____

   _____

   _____

2. What are the potential limitations of the Unicode chart used in this activity?

   _____

   _____

   _____

Using Non-Keyboard Characters

There are more character codes than can fit on a keyboard. It is possible to use these codes in your output statement. In this activity, you will insert non-keyboard characters into a System.out.println( ) statement. To complete this activity, add this option to the Run Flags2 for jGRASP:

   -Dfile.encoding=UTF-8.

Procedure

1. Scan through pages 3 to 6 in the code chart used in the previous activities. Locate code 0021, which is an exclamation mark (!). Look at the related codes in that section. The code for a double exclamation mark is 203C. If you add that at the end of your name in the UnicodeName.java file, jGRASP will print a double exclamation mark after your name. For example:

   ```
   System.out.println( "\u004C\u0061\u006B\u0065\u0073\u0069\u0061\u203C" );
   ```

   The output is:

   ```
   Lakesia‼
   ```

   Browse the Unicode chart and record codes you might like to incorporate into print statements in your program.

   _____

   _____

2. Select a couple of the codes you identified in the previous question. Write several print statements to tell a story. Outline your plan in the space that follows and then write your new program. It seems silly to use Unicode when characters on the keyboard are required. Use the keyboard characters when available and add the non-keyboard characters in Unicode as needed. For example:

   ```
   System.out.println( "Lakesia\u203C" );
   ```

   The output is:

   ```
   Lakesia‼
   ```

   In the UnicodeName.java file, change the class name to NonKeyboardCharacters, and save the file as NonKeyboardCharacters.java in your working folder.

   _____

   _____

3. In your web browser, return to the Code Charts page on the Unicode website. Open the Latin-1 Supplement chart. This chart contains accented and otherwise marked alphabetic characters that may be a part of your name. If so, record the codes below. If there are no such characters in your name, record the codes for the Spanish word for tomorrow: mañana. Use the keyboard characters when available and add the non-keyboard characters as needed. Add these codes to a print statement in your program.

   _____

   _____

Reflections

1. Why do you think Unicode supports so many variations of the alphabet?

   _____

   _____

2. Obviously there is not enough room on the keyboard to include all Unicode characters. What codes would you add to the keyboard?

_____
_____
_____

Debugging Challenge

Apply what you have learned about Java program structure to find six errors in the following code. Classify the errors as compiler, logic, runtime, or user. Record your answers in the space that follows.

```
/* Lab Manual Chapter 3 - Debugging Challenge
   your name here
*/
public class Debugging Challenge {
   public static void Main( String [ ] args ) [

      /* output your name.*/
      System.out.printline( "My naem is Darlene." ):

   ] // end main
} // end class
```

_____
_____
_____
_____
_____

To verify you found all the errors, launch jGRASP, and open the DebuggingChallenge.java file downloaded from the student companion website. Enter the corrections you identified. Then, compile and run the program.

Lab 3-2

Understanding Errors

Programming without errors is certainly the goal of great programmers. However, while you are in the learning phase, errors will occur. The jGRASP IDE notifies coders of compiler and runtime errors. Logic errors will require close examination and testing by the programmer. In this lab, you will explore the error-reporting features of jGRASP.

Learning Goals
- Describe error reporting features of the jGRASP IDE.

Materials
- jGRASP Integrated Development Environment
- Starter files from the student companion website
- Appendix C Responding to Java Error Messages in the textbook

Application and Extension of Knowledge

Exploring jGRASP

Your first Java program was written in jGRASP, the Integrated Development Environment. If no typos were made, the program ran with a clean compile the first time. To observe how jGRASP handles errors, you will load a Java file containing errors and interact with jGRASP.

Procedure

1. Examine this code. Make note of errors you can readily identify in the space below.

```
1   /* Lab Manual Chapter 3 - Error Reporting in jGRASP
2     your name here
3
4   public class ErrorReporting {
5     public static void main( String [ ] args ) (
6
7       /* Report the temperature using Unicode for the degree symbol.*/
8       SystemOutPrintln( "The temperture is 76\u 2070." );
9
10    ) // end main
11  } // end class
```

_____
_____
_____
_____
_____

2. Launch jGRASP, and open the ErrorReporting.java file downloaded from the student companion website.

3. If line numbers are not displayed, click **View>Line Numbers** so it is checked. The [Ctrl][L] key combination can also be used to toggle the line numbers on and off.

4. Click the **Compile file** button (green plus sign) to compile the file. Make note of the three error messages. Do your best to figure out what is wrong. The carets (^) below the lines of code may offer a hint. This is how jGRASP indicates where in the line the compiler found the error. The compiler's best guess at the type of error is in the first line of the error message. Briefly describe what you think is wrong.

_____
_____
_____
_____
_____

5. The first error reported is due to a missing class. However, there is a class. It is ErrorReporting. The cause of this error is that there is no */ to end the comment. Because the compiler was instructed to ignore comments, it never saw the class definition. As written, the */ on line 7 ends the comment started on line 1. Thus, line 8 was the first line the compiler tried to compile. Add the */ on line 3. Notice the color change in the code. Use the colors to help you debug your code.

Name _____   Chapter 3 Introduction to Java Programming   **31**

6. Click the **Clear** button on the **Compile Message** tab next to the error messages. This clears the messages in the **Compile Message** tab.

7. Recompile the file. Now there are three errors! Here is what happened. On the first compile, jGRASP got as far as it could and then reported the errors. On the second compile, with the first error repaired, jGRASP could get further and found three more errors. Notice that the second error is one you saw on the first compile. Make a note of the other two messages in the space below.

   _____

   _____

   _____

8. Do your best to figure out what is wrong. The error message may not be as helpful this time. Apply what you know about the main method to identify a typo in the punctuation. Write your findings below.

   _____

   _____

   _____

9. Repair the code by replacing both the beginning parenthesis at the end of line 5 and the ending parenthesis on line 10 with braces.

10. Clear the error messages and recompile. Only one error message remains. Find the error indicated by the error message "illegal unicode escape" and the caret below the Unicode. What is the error?

    _____

    _____

    _____

11. Remove the extra space in the Unicode.

12. Clear the error messages and recompile. What is the result? Enter the result below.

    _____

    _____

    _____

13. Repair the typos in the System.out.println command.

14. Clear the error messages and recompile. Enter the result below.

    _____

    _____

    _____

Copyright Goodheart-Willcox Co., Inc. May not be reproduced or posted to a publicly accessible website.

15. You should have a clean compile with no more compiler errors. Click the **Find and run main method or applet** button (red runner) to run the program. Enter the result below. Are there any more errors?

    _____
    _____
    _____

16. There is one remaining error the compiler did not flag. This is a logic error. Correct the spelling of *temperature,* compile, and run the program again. Enter the result below.

    _____
    _____
    _____

Reflections

1. Discuss the value of careful input of the code the first time.

    _____
    _____
    _____

2. Explain why you think the compiler does not just fix the code to correct the errors it finds.

    _____
    _____
    _____

Debugging Challenge

You write code, compile it, and run it. It has no compile, runtime, or logic errors. You send the source code to another coder (your user) who writes back that it will not run. The user reports that it has no output. You scratch your head. You ask the user how he or she is trying to run the program. The reply states that the user is clicking the green plus sign button in jGRASP. When doing this, the code compiles, but does not run. What is the error?

_____
_____
_____
_____
_____
_____
_____
_____

Name _____ Date _____ Class _____

CHAPTER 4

Variables

In this chapter, you learned about data types and identifier creation. You are now able to store numbers and letters in memory locations. This lab will help you to understand why data types are necessary.

Chapter Highlights

- Named data can be stored in memory while the program executes so the value can be retrieved as needed.
- Naming data allows a generalized algorithm to be created that works with any data.
- A *variable* is a named memory location defined by name and data type whose value is able to be changed during program execution.
- *Identifiers* are names the programmer chooses for the variables in programs. They are also used as names for programs and for any methods the program defines.
- A name must start with a letter, underscore, or dollar sign. Numbers are allowed after the first character, but a name cannot start with a number.
- Spaces are not allowed in names and names are case-sensitive.
- Java's keywords and reserved words cannot be used as names.
- The *data type* is the format in which the data will be stored and the size the variable will be given in memory.
- Four of the primitive data types that can hold integers are byte, short, int, and long. The float and double data types hold real numbers.
- The char data type holds one Unicode character.
- The boolean data type can be either of the reserved words true or false.
- A *literal value* is simply a textual representation of a value.

While studying, look for the activity icon for:

- Vocabulary terms with e-flash cards and matching activities.
- Starter files for lab activities.

These activities can be accessed at
www.g-wlearning.com/informationtechnology/1773

- To assign a value directly to a variable, use the assignment operator (=).
- A *constant* is a value that cannot and should not change during the execution of the program. To define a constant, use the keyword final in front of the definition.
- To input values from the user, the Scanner class is used.

Warm-Up Exercises

_____ 1. Consider the following code segment. What is printed as a result of executing this code?

```
int alpha = 4;
double beta = 15.5;
long gamma = alpha;
beta = gamma;
System.out.println( beta );
```

A. 15.5

B. 4.0

C. 4

D. Error: Type mismatch

E. beta

_____ 2. Consider the following code segment. What variable storage is allocated as a result of executing this code segment?

```
int alpha = 4;
double beta = 15.5;
long gamma = alpha;
beta = gamma;
System.out.println( beta );
```

A. 17 bytes (alpha: 1 byte; beta: 8 bytes; gamma: 8 bytes)

B. 20 bytes (alpha: 4 bytes; beta: 8 bytes; gamma: 8 bytes)

C. 12 bytes (alpha: 4 bytes; beta: 4 bytes; gamma: 4 bytes)

D. 6 bytes (alpha: 2 bytes; beta: 2 bytes; gamma: 2 bytes)

E. 8 bytes (only beta is actually stored)

_____ 3. Consider the following code segment. Which of the variable definitions are allowed?

```
int alpha;
double _beta&;
long 1A;
```

A. alpha and _beta&

B. alpha only

C. alpha and 1A

D. 1A and _beta&

E. _beta& only

_____ 4. What is the difference, if any, between System.out.println( ) and System.out.print( )?
   A. System.out.println( ) prints what is in the parentheses and goes to the next line, while System.out.print( ) prints what is in the parentheses and stays on the same line.
   B. System.out.println( ) is used for string literals only, while System.out.print ( ) is used for numerical data.
   C. System.out.println( ) and System.out.print( ) can be used for user input.
   D. System.out.print( ) prints what is in the parentheses and goes to the next line, while System.out.println( ) prints what is in the parentheses and stays on the same line.
   E. The two commands are identical.

_____ 5. Consider the following code segment. What is printed as a result of executing this code?
```
char gradeA = 'A';
char gradeB = 'B';
System.out.println( "I got an " + gradeA + " or a " + gradeB );
```
   A. I got an gradeA or a gradeB
   B. I got an A or a B
   C. A
   D. B
   E. A or B

_____ 6. Consider the following list of identifiers. Which one is a valid definition?
   A. int static;
   B. double true;
   C. long void;
   D. float boat;
   E. short class;

_____ 7. A variable is needed to store the population of the world. That number exceeds seven billion individuals. What data type should be used?
   A. int
   B. double
   C. long
   D. float
   E. char

_____ 8. A variable is needed to hold the national debt. The debt is in the neighborhood of 22 trillion dollars. What data type should be used?
   A. float or double
   B. double or int
   C. long or short
   D. float or char
   E. boolean

_____ 9. You are coding a card game for a mobile app. You want to display the suit symbols: ♣ ♦ ♥ ♠. Fortunately, these are found in Unicode, but the color is black. Your app can add formatting to color the diamond and heart red. What data type should be used for these symbols?
   A. char
   B. byte
   C. int
   D. double
   E. boolean

_____ 10. You are coding a quiz for your classmates to practice their Java vocabulary terms. You want to store whether or not they have the right answer. What data type should be used?
   A. char
   B. byte
   C. int
   D. double
   E. boolean

Lab 4-1
Identifiers and Data Types

The process of creating identifiers is very structured in Java. In this lab, you will apply the rules for creating identifiers and develop skill with CamelCase. Different numbers of bytes are allocated for the various Java data types. You will also apply what you learned about binary and data storage to find the maximum and minimum values that can be stored in each data type. Before beginning this activity, download the files for this lab from the student companion website.

Learning Goals
- Select identifiers for data.
- Identify the correct data type.
- Illustrate the process for defining variables in Java.

Materials
- Handheld or computer-based calculator
- jGRASP Integrated Development Environment
- Starter files from the student companion website

Application and Extension of Knowledge

Naming Variables

The names of variables in Java serve several purposes. These identifiers stand in for the values the variables hold. They communicate to the programmers the meaning of the data held inside. To do this, rules must be followed for creation of identifiers. Sadly, the data type is rarely available from the variable name. In an early language, FORTRAN, all variables starting with the letters from I to N were integers. All other variables held floating-point numbers. Java does not do that. Consequently, it is up to you to keep track of the data types or else mixed-case errors may occur. This activity applies what you have learned about the rules for naming variables and demonstrates how identifiers are important in programming.

Procedure

1. List the rules for naming variables in Java. For a refresher, consult the chart in Figure 4-1 in the textbook.

2. The process of writing a variable definition results in a Java statement. The three steps are (1) name the data type, (2) write the identifier, and (3) append a semicolon. Assume you need a variable to hold a temperature. Write a Java statement to define this variable and explain why you chose this data type.

3. In addition to the rules for identifier creation, there are style suggestions. Consider a variable name that follows the rules, but is not easy to read:

   ```
   int ageinyears;
   double ageinfractionsofyears;
   ```

   It is helpful to apply CamelCase to variable names. In CamelCase, the words that describe the content of the variable are written without spaces. To help the reader find the individual words in the identifier, the name starts with a lowercase letter, and the first letter of each succeeding word is capitalized. For example: heightininches is written as heightInInches in CamelCase. It is a generally accepted style to start a variable identifier with a lowercase letter. Write the two variable definitions above using CamelCase.

4. Keeping track of your personal finances is very important. Several key values are vital. Write definition statements for the following variables. Apply the rules for creating Java identifiers and CamelCase.

   A. savings account balance _____
   B. checking account balance _____
   C. car loan balance _____
   D. savings account interest rate _____
   E. car loan interest rate _____
   F. months left on car loan _____
   G. bills paid (true or false) _____

5. Launch jGRASP, and open the IdentifiersSnippet.java file. Check your answers for the previous question by writing your statements into the file. Compile and verify that these are legitimate variable names. For any that generate compiler errors, fix the names and compile again. Enter any errors and describe how you fixed them in the space provided. Then, update the definitions in the previous question to match the final results.

_____

_____

_____

Reflections

1. Explain why there are rules for creating variable names in Java.

_____

_____

_____

2. How do naming conventions help you and future programmers who will edit your work?

_____

_____

_____

Identifying Maximum and Minimum Values in Data Types

The maximum and minimum values an integer variable can hold depend on the number of bytes allocated to the variable. This determination is made by the selection of a data type. It is a straightforward calculation based on the binary number system. In this activity, you will find the maximum and minimum values for the integer data types.

Procedure

1. Use a calculator or practice your mental math to find the powers of 2 from 0 to 16.

Exponent	Power of 2
0	
1	
2	
3	
4	
5	
6	
7	
8	

Exponent	Power of 2
9	
10	
11	
12	
13	
14	
15	
16	

2. The largest number that four binary digits (bits) can hold is 1111. Write the value of this binary number in decimal.

3. The fourth power of 2 is 16. You would expect four bits to hold 16 values. Well, it does—the values are 0 to 15. Write the numbers from 0 to 15 and count them. How many numbers are there?

4. The integer data type **byte** is allocated 1 byte of storage. Repeat the analysis done in the previous question to find the range of positive numbers a byte can hold.

5. In Java, one bit is reserved for a positive (0) or negative (1) sign, so there are only seven bits left in a byte to represent a positive number.

   A. What is the largest positive number a byte can hold if the first bit is reserved for a sign?

   B. Write the largest positive binary number in a byte.

   C. Write the smallest negative binary number in a byte.

   D. Write its decimal equivalent.

6. You found that a **byte** data type can hold eight binary digits and 256 different numbers. The range of those numbers is −128 to +127. Perform the same analysis on a **short** data type.

   A. Find the largest positive number.

   B. Find the smallest negative number.

   C. Give the range of numbers a **short** can hold.

Reflections

1. Explain why Java has a limit on the sizes of the values stored in integer data types.

2. In early programming languages, there was no to very little data typing. No checking went on to see that the code has not stored a number too big for the memory location allocated. If the number were too big, the higher digits were simply stored in the next byte over. What do you think this would mean for what was already stored in the next byte over?

Debugging Challenge

Consider the following code. List the error messages you might get when attempting to compile in the space provided. Once you have decided about the validity of each definition, launch jGRASP, open the Debug4_1.java file, and verify your answers by modifying and compiling the file.

```
1  /* Debugging Challenges Chapter 4 */
2
3  public class Debug4_1 {
4      public static void main( String [ ] args ) {
5
6          // define variables
7          byte bite;
8          short _tiny;
9          int integer-5;
10         long 9_quintillion;
11         float b0at;
12         double B0at;
13         char unicode;
14         boolean true;
15
16     } // end main
17 } // end class
```

Lab 4-2

Variable Values

Variables are not very useful until they hold information for the program. In most programs, variables have an initial value. That value may change during program execution. In some cases, the value should not change; these should be defined as constant values. Other values can be input by the user. In this lab, you will assign values to variables and create constants. This set of activities helps you make assignments using Java statements.

Learning Goals
- Assign values to variables by using Java statements.
- Assign values to constants.
- Assign values to variables by getting input from the user.

Materials
- jGRASP Integrated Development Environment
- Starter files from the student companion website

Application and Extension of Knowledge

Writing Assignment Statements

In Java, values are stored into a variable using the assignment operator (=). The variable is the first item in the Java statement, followed by the equals sign, and ending with the value. Care must be taken that the values are of a data type compatible with the data type of the variable. You can assign a variable of lower precision to a variable of higher precision. For example:

```
double a = 10;
```

Procedure

1. Write the Java statements to assign the given values when defining variables for these items: length of a board (5.25'), length of shelf (26"), and number of boards on hand (12).

   _____
   _____
   _____

2. Launch jGRASP, and open the AssignmentStatementsSnippet.java file. Examine the code. Write down what you anticipate the output would be.

   _____
   _____
   _____

3. Compile the snippet, and write down the outcome.

   _____
   _____
   _____

4. Compiling the file in the previous question may have resulted in an unexpected error. Often while coding, assumptions are made that the compiler cannot deal with. You may get a perplexing error message and must figure out what went wrong. A problem arises if you try to output the inches sign ("). It is the same as the String literal symbol. Putting it into the statement generates an error due to the mismatched number of quotation marks. There are three quotation marks in the statement. Perhaps you noticed the error before compiling because of the inconsistent coloring of the code. It is a good idea to take clues from the color in the editor. There is a mechanism in Java that allows a fix for needing a quotation mark inside a String. It is called an *escape sequence.* The backslash (\) is an escape symbol. It tells the compiler to escape out of the String, ignore the quotation marks, and print the actual inch symbol (quotation mark). For example, if a statement is:

```
System.out.println( "The length is 5\"." );
```

then the output is:

```
The length is 5".
```

Add the \" to the output statement, recompile the file, and run the program. Record your output in the space provided.

_____

_____

_____

5. Continue working in the file, and enter the statements you wrote in question number 1. Generate the control structure diagram (**View>Generate CSD**), and then remove the CSD (**View>Remove CSD**) to ensure all statements are properly aligned. Compile the file. Fix any errors. Then, run the program. Make a statement about what you anticipated the output would be and what the actual output was.

_____

_____

_____

Reflections

1. Explain why the escape sequence \" is required to print the inch symbol.

_____

_____

_____

2. Describe what happens when the CSD is generated, and what happens when the CSD is removed. Why is this sequence of steps performed?

_____

_____

_____

Name _____

Defining Constant Values

Variables usually have an initial value, which may change during program execution. For some variables, the value does not change for the entire execution of the program. It is better to define these values as a constant. This activity helps you define constants using Java statements.

Procedure

1. Launch jGRASP, and open the DefineConstantsSnippet.java file. Examine the code. What is the keyword that indicates a constant is being defined?
   _____

2. Compile and run the snippet. Enter the output in the space provided.
   _____

3. Locate comment 3 in the code, enter statements that assign INCHES_IN_FOOT to an int variable identified as foot. Be sure to define the int before using it. Print the new value of foot. Enter the output in the space provided.
   _____
   _____

4. The word *constant* means the same forever. Now, see what happens when you try to reassign a constant. Locate comment 4 in the code, and enter this statement:
   ```
   INCHES_IN_FOOT = 10;
   ```
   Recompile the code. Enter the compiler output in the space provided.
   _____
   _____
   _____

Reflections

1. Explain what "magic numbers" are and how they can make coding, and especially editing code, difficult.
   _____
   _____
   _____

2. Explain how you would handle a situation in which you thought there might be a need to change a constant value during execution.
   _____
   _____
   _____

Assigning Values from the User

Variables can be assigned an initial value and then modified by program execution. In every interactive program, the user enters new information that is used in the calculations. The Scanner class provides user input in Java programs.

Procedure

1. In the space provided, write the import statement required to use Scanner in a program.
   _____

2. In the space provided, enter an instantiation statement to use the Scanner class via an object named input.
   _____

3. Launch jGRASP, and open the ValuesFromUserSnippet.java file. This program asks the user for profile information and prints it for verification. The first prompt and user input as well as the echo of the information back to the user is provided for you. In the space provided, explain the purpose of the statement on line 13.
   _____
   _____
   _____

4. Locate comment 1. Write a prompt statement to ask the user for his or her age. Define an appropriate variable and use Scanner to accept the user response. In the space provided, try to anticipate what could go wrong during user input.
   _____
   _____
   _____

5. How might you be able to avoid an input error with clear instructions in the prompt?
   _____
   _____
   _____

6. Continue completing the code for comment 1. Name the data types that may be used for each profile data point variable. It is possible more than one data type may be appropriate for each variable.

   A. id _____

   B. age _____

   C. grade _____

   D. gpa _____

   E. number of siblings_____

   F. number of pets_____

Name _____

7. Compile and run the program. Enter any errors and describe how you fixed them in the space provided. It is a good memory aid to reflect on the errors. Writing down the errors and what was required to correct them helps you remember to avoid these mistakes later.

_____
_____
_____

8. Locate comment 2. *Echo* is a programming word for saying back to the user what the user has entered. Add code to echo the user input with an appropriate label. Compile and run the program. Enter the errors encountered and describe how you fixed them in the space provided.

_____
_____
_____
_____
_____

9. Locate comment 3. Add code to ask the user if the information printed is correct. Because characters cannot be accepted, use the convention that 1 = yes, and 0 = no. Echo what the user entered. In later chapters, you will learn how to act on the response. For today, simply get the answer and echo it. Compile and run the program.

10. In your testing, enter inappropriate responses. Note the reaction from the Java runtime engine. These are runtime errors, caused by user errors. In the space provided, enter one thing you learned in this lab.

_____
_____
_____

Reflections

1. Explain the purpose of the Java Class Library.

_____
_____
_____

2. Describe ways to avoid user errors.

_____
_____
_____

Debugging Challenge

Determine the error in this code and propose a fix in the space provided.

```
/* Debugging Challenge Chapter 4 */

import java.util.Scanner; // this imports the Scanner class

public class Debug4_2 {
   public static void main( String [ ] args ) {
      Scanner input = new Scanner( System.in ); // create the input object

      /***** 1. prompt the user for the number of books they have read
                 this year, read the value into a variable of the appropriate
                 type, and output the value in a message
      */
      System.out.print( "How many books have you read this year? " );
      int booksRead = input.nextDouble( );

      System.out.println( booksRead + " is a lot of books!" );

   } //end main
} // end class
```

Name _____ Date _____ Class _____

CHAPTER 5

Java Expressions

Expressions describe how the values are combined to get a new value. Programmers create variables and write expressions to calculate results using constants and values stored in variables.

Chapter Highlights

- Java provides these arithmetic operators: + (addition), − (subtraction), * (multiplication), / (division), and % (modulus).
- Java evaluates expressions by following this order of precedence: parentheses; then multiplication, division, and modulus; then addition and subtraction; then assignment.
- Java also provides these shortcut operators: ++ (shortcut increment operator); -- (shortcut decrement operator); and +=, −=, *=, /=, %= (shortcut arithmetic operators).
- Division involving two integers results in an integer. Division involving at least one floating-point value results in a floating-point value.
- Values can be cast to a different data type by preceding the value with the new data type in parentheses.
- The result of dividing by zero depends on the data types. If the numerator and denominator are both integers, an exception occurs. If the numerator is a floating-point number other than 0.0, the result is infinity. If the numerator is 0.0, the result is NaN (not a number).
- Numbers can be formatted for output using the System.format( ) method. The format specifier can define the data type, width, alignment, and precision of the outputted value.

 While studying, look for the activity icon for:

- Vocabulary terms with e-flash cards and matching activities.
- Starter files for lab activities.

These activities can be accessed at
www.g-wlearning.com/informationtechnology/1773

Warm-Up Exercises

1. Assume the following variables have been declared:

   ```
   int i;
   double g;
   ```

   In the space provided, write the value that will be assigned to the variable after each of the following statements are executed.

   A. i = 9 / 4; _____

   B. i = 9 % 4; _____

   C. g = 9.0 / 4.0; _____

   D. i = 12 / 6 * 3; _____

   E. i = 12 / (6 * 3); _____

   F. i = 6 + 6 / 2; _____

   G. g = (double) (7) / 5; _____

   H. g = (double) (7 / 5); _____

   I. i = 7;
      i *= 5; _____

   J. g = 0.0 / 0.0; _____

_____ 2. Consider the following code segment. What is printed as a result of executing this code segment?

   ```
   int d = 4;
   int c = 15;
   int b = c - d;
   int a = c % b;
   System.out.println( a );
   ```

   A. 11

   B. 4

   C. c % d

   D. 15.4

   E. a

_____ 3. Consider the following code segment. What is printed as a result of executing this code segment?

   ```
   final double LENGTH = 4.25;
   final double WIDTH = 16;

   double area = LENGTH * WIDTH;
   System.out.println( "Area is " + area );
   ```

   A. 68

   B. 68.0

   C. Error: Type mismatch

   D. Area is 68.0

   E. Area is 64.0

_____ 4. Consider the following code segment. What is printed as a result of executing this code segment?

```
int peopleInVan = 5;
final int MAX_SEATING_IN_VAN = 8;

peopleInVan -= 2;
peopleInVan *= 2;
int spacesInVan = MAX_SEATING_IN_VAN - peopleInVan;
System.out.println( spacesInVan );
```

A. 8
B. 8.1
C. 6.0
D. 4
E. 2

_____ 5. Consider the following code segment. What is printed as a result of executing this code segment?

```
double a = 6.0;
int o = 42;
int p = 24;

int m = o / p;
int n = o % p;

double q = a + m - n;
System.out.println( q );
```

A. −11.0
B. 11.0
C. −12.5
D. 11
E. 0

_____ 6. Consider the following variable definition. What is the value of x?

```
double x = 2.5 * 3 - 5 * ( 28.0 / 4 + 3 );
```

A. −42.5
B. 10.0
C. −12.5
D. 25
E. 24.5

7. Letitia owns a smoothie truck in Mississippi that specializes in making smoothies with organic fruits. She has two part-time employees who each earn $30,000 per year. Her fixed operating costs are $244 per month. Her average cost to make a smoothie is $2.70, and she sells them for $6.50. To compute how many smoothies the truck must sell in a month to break even, Letitia must divide total fixed costs by profit per smoothie. Write a Java expression that will calculate her break-even number of smoothies per month.

_____

_____

8. The three key quantities in a business are assets, liabilities, and net worth. Assets are things that the business owns such as furniture. Liabilities are monies owed to creditors. Net worth is the value of the business. The basic accounting equation is:

   net worth = assets – liabilities.

   Write a code segment that defines these three values. Then, write three equations to calculate net worth, assets, or liabilities. Test it with these values: netWorth = 100000.00; assets = 300000.00; and liabilities = 200000.00.

9. Four double values determine the coordinates of a line segment. Write a code segment that uses values in the following variables to calculate the slope of a line segment. The formula for slope is:

   $$\text{slope} = \frac{y1 - y2}{x1 - x2}$$

   ```
   double x1 = 3.0;
   double y1 = 6.0;
   double x2 = -4.0;
   double y2 = 10.0;
   double slope;
   ```

10. The formula for the surface area of a box is:

    surface area = $2lw + 2wh + 2lh$

    where l is length, w is width, and h is height.

    Define three double variables and assign values. Define a double variable surfaceArea, and assign the result of the formula calculation.

Lab 5-1

Arithmetic Operators

In this lab, you will apply the rules for arithmetic operations by solving problems requiring calculations. Before beginning this activity, download the files for this lab from the student companion website.

Learning Goals
- Apply binary operators in Java.
- Construct a Java expression with proper operator precedence.
- Use a shortcut operator in Java code.

Materials
- jGRASP Integrated Development Environment
- Starter files from the student companion website

Application and Extension of Knowledge

Calculating Numbers of Outfits

Have you ever wondered how many potential outfits you have in your closet? In this activity, you will compute the number of unique combinations you have in your wardrobe.

Consider you have six tee shirts and four pairs of jeans in your wardrobe. To calculate the number of possible outfits you can make with these items, multiply 6 times 4. There are 24 possible different outfits with six tee shirts and four pairs of jeans. Next, consider you also have three hooded sweatshirts. How many possible outfits can you make with six tee shirts, four pairs of jeans, and three hooded sweatshirts?

___

___

___

Begin with the ClothingSnippet.java file. Edit it to accept input from the user for the number of tee shirts, jeans, and sweatshirts. Calculate the number of possible outfits and report the result to the user.

Procedure

1. Follow this algorithm, and plan your solution in the space provided. Write generalized statements that will help you design the code later.

    ```
    /***** 1. Get the items of clothing from the user.
            Prompt the user for the number of tee shirts.
            Prompt the user for the number of jeans.
            Prompt the user for the number of hooded sweatshirts.
    */
    ```

___

___

___

___

```
/***** 2. Calculate the number of possible different outfits. */
```

_____
_____
_____

```
/***** 3. Output the result. */
```

_____
_____
_____

2. Launch jGRASP, and open ClothingSnippet.java starter file.
3. Enter code for the generalized statements you wrote above.
4. Test the program, and correct any errors as needed.

Reflections

1. Explain whether this method will always produce a wearable outfit.

_____
_____
_____

2. Suppose footwear is considered as part of an outfit. How would the algorithm and the code change?

_____
_____
_____

Calculating a Grade Average

You are taking a course with an unusual grading scheme. You would like to know your current grade in the course. Consider there are two tests and five quizzes in a quarter. The average of the quizzes counts the same as a test. The tests and quizzes are scored numerically from 0 to 100, and the score may contain decimals.

Begin with the AveragesSnippet.java file. Define variables for the two tests and the five quizzes, and assign values to the variables. Calculate the average of the quizzes. Find the average of the three scores (test1, test2, and quizAvg).

Procedure

1. Follow this algorithm, and plan your solution in the space provided. Write generalized statements that will help you design the code later.

```
/***** 1. Assign the two test scores to double variables test1 and test2.
*/
```

_____
_____
_____

```
/***** 2. Assign the five quiz scores to double variables quiz1,
          quiz2, quiz3, quiz4, and quiz5.
*/
```

_____

_____

_____

```
/***** 3. Assign the average of the quizzes to quizAvg. Hint: use the += operator to
          add the quizzes to quizAvg. This makes it easier to modify the code if the
          number of quizzes changes. Keeping in mind casting and promoting, divide
          quizAvg by 5.
*/
```

_____

_____

_____

```
/***** 4. Add the test1, test2, and quizAvg. Then, divide by 3. Store the result in
          grade. */
```

_____

_____

_____

```
/***** 5. Output the result. */
```

_____

_____

_____

2. Prepare a test case for this algorithm. Assign numbers to the tests and quizzes, and hand-check the average in the space provided.

_____

_____

_____

_____

3. Launch jGRASP, and open **AveragesSnippet.java** starter file.
4. Enter code for the generalized statements you wrote above.
5. Test the program, and correct any errors as needed.

Reflections

1. Why do you think the teacher devised this grading scheme?

_____

_____

_____

2. How would this algorithm change if the grades were letter grades?

___

___

___

___

Verifying Euler's Formula

In mathematics, solid figures or polyhedrons are three-dimensional objects that have polygons for faces. The word *polyhedron* originates from a Greek word meaning "many faces." The great 18th-century Swiss mathematician Leonhard Euler found that the number of vertices (corners) minus the number of edges plus the number of faces is always 2. Euler's polyhedral formula is:

$$V - E + F = 2$$

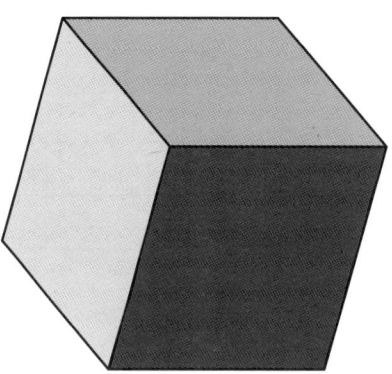

Goodheart-Willcox Publisher

Consider the cube in the figure. It has square faces. Mentally count the number of vertices or corners on the cube. There are ____ vertices. Count the number of edges. There are ____ edges. Count the number of faces. There are ____ faces. Subtract the edges from the vertices, then add the faces. The result is ____.

Begin with the EulerSnippet.java file. Modify this program to verify Euler's formula for cubes.

Procedure

1. Follow this algorithm, and plan your solution in the space provided. Write generalized statements that will help you design the code later.

   ```
   /***** 1. Define vertices, edges, and faces for a cube and assign values to variables. */
   ```

   ___

   ___

   ___

   ___

   ```
   /***** 2. Apply arithmetic operators to calculate Euler's polyhedral formula. */
   ```

   ___

   ___

   ___

   ___

Name _____   Chapter 5 Java Expressions   55

```
/***** 3. Output the result. */
```
_____
_____
_____
_____
_____

2. Launch jGRASP, and open EulerSnippet.java starter file.
3. Enter code for the generalized statements you wrote above.
4. Test the program, and correct any errors as needed.

Reflections

1. Think about a tetrahedron. It has four triangular faces. Verify Euler's formula for a tetrahedron: four faces, six edges, and four vertices.
_____
_____
_____
_____
_____
_____

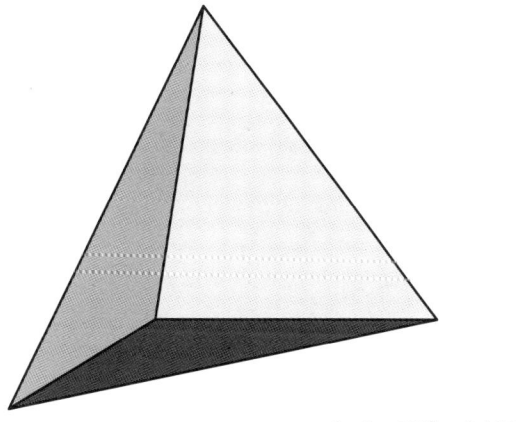

Goodheart-Willcox Publisher

2. Explain how you might be able to calculate the number of vertices in a polyhedron if you know the number of faces and edges.
_____
_____
_____
_____

Debugging Challenge

You are making favors for a party. You will need two favors for each party guest. You expect 20 guests to attend. There are five days before the party, and you have completed 10 favors so far. How many favors do you need to make per day to be ready for the party? You try this statement to calculate the answer:

```
favorsPerDay = 20 * 2 - 10 / 5;
```

The result from this statement is 38. You know this is not correct since you need to make 40 favors total and you have already made 10. What is wrong with the code, and what is the correct answer? Record your answers in the space provided.

_____

_____

_____

_____

_____

Lab 5-2

Operators and Expressions

In this lab, you will apply the rules for arithmetic operations by solving problems requiring expressions. Before beginning this activity, download the files for this lab from the student companion website.

Learning Goals
- Apply casting and promotion in mixed-type arithmetic.
- Select the proper data type for the information to be stored.

Materials
- jGRASP Integrated Development Environment
- Starter files from the student companion website

Application and Extension of Knowledge

Calculating Constant Acceleration

Rockets to take humans and cargo to space accelerate at a steady rate. With constant acceleration, the speed of the spaceship will change. With a positive acceleration, the spaceship will go faster and faster. A formula for calculating the speed of an object traveling with constant acceleration is:

speed = initial speed + acceleration × time

This activity provides a chance to work with an Earth-bound body with constant acceleration. Begin with the AccelerationSnippet.java file. Edit the program to prompt for the initial speed, the number of seconds, and the acceleration during those seconds. Then, calculate the final speed. Output the data entered and the final speed.

Name _____ Chapter 5 Java Expressions 57

Procedure

1. Follow this algorithm, and plan your solution in the space provided. Write generalized statements that will help you design the code later.

   ```
   /***** 1. Get the data from the user.
             Prompt the user for the initial speed in meters per second.
             Prompt the user for the acceleration in meters per second squared.
             Prompt the user for time in seconds.
             Select proper data types for the information to be stored.
             Use meaningful variable names.
   */
   ```

   _____
   _____
   _____

   ```
   /***** 2. Calculate the final speed. */
   ```

   _____
   _____
   _____

   ```
   /***** 3. Output the result. */
   ```

   _____
   _____
   _____

2. Launch jGRASP, and open **AccelerationSnippet.java** starter file.
3. Enter the code you wrote above. Enter code for the generalized statements you wrote above.
4. Test the program, and correct any errors as needed.

Reflections

1. What can you expect to happen if the user enters zero or a negative number for the acceleration?

   _____
   _____

2. Explain your choice of data type.

   _____
   _____
   _____

Finding Force Due to Gravity

Sir Isaac Newton is credited with the discovery of gravity. He published the Law of Universal Gravitation in the 17th century. According to Newton, any two objects attract each other equally with a gravitational pull proportional to the product of their masses and inversely proportional to square of the distance between their centers of mass. The force of the pull is calculated using this formula:

$$F = G \frac{(massObject1)(massObject2)}{distance}$$

where G is the gravitational constant 6.674×10^{-11}.

You will write a program to calculate the force for two objects. Edit the GravitySnippet.java file to accept the masses and their distance as input, calculate the force of attraction, and output the results.

Procedure

1. Follow this algorithm, and plan your solution in the space provided. Write generalized statements that will help you design the code later.

   ```
   /***** 1. Capture the masses of the two bodies and the distance between them.
            Prompt the user for the mass of the first body.
            Prompt the user for the mass of the second body.
            Prompt the user for distance between the two bodies.
   */
   ```

   ```
   /***** 2. Calculate the gravitational pull between the bodies. */
   ```

   ```
   /***** 3. Output the result. */
   ```

2. Launch jGRASP, and open GravitySnippet.java starter file.
3. Enter code for the generalized statements you wrote above.
4. Test the program, and correct any errors as needed.

Reflections

1. Based on the output from your program, what can you say about the size of the pull from gravity?

2. Look up the mass of Earth. Enter your own mass, or weight. Estimate the distance from the center of the Earth to your center of mass. Predict what the pull will be. Explain your prediction. Compare it with the output. Write a statement about the comparison.

How Fast Are You Running?

You are running a five-kilometer (5K) race. You want to know your total time in minutes and seconds to run the race. You also want to know your average time per kilometer. Write a program to input your start and end times in hours, minutes, and seconds. Hint: convert the times into seconds since midnight. Use RaceTimeSnippet.java as a starter file.

Procedure

1. Follow this algorithm, and plan your solution in the space provided. Write generalized statements that will help you design the code later.

```
/***** 1. Prompt the user for the start time in hours and minutes and seconds.
          Prompt the user for the end time in hours and minutes and seconds.
          Use meaningful variable names.
*/
```

```
/***** 2. Convert the times to seconds. */
```

```
/***** 3. Subtract the end time from the start time, then convert back to hours,
          minutes, and seconds
*/
```

```
/***** 4. Output the total time. */
```

```
/***** 5. Calculate average time per kilometer in seconds. */
```

```
/***** 6. Output the average time in seconds. */
```

_____
_____
_____

2. Launch jGRASP, and open RaceTimeSnippet.java starter file.
3. Enter code for the generalized statements you wrote above.
4. Test the program, and correct any errors as needed.

Reflections

1. Consider the manner of inputting the times. Compare that with the way mobile apps request a time.

_____
_____
_____

2. What would need to change in the code to output the average time per kilometer in minutes and seconds?

_____
_____
_____

Debugging Challenge

You are calculating the average age of your family members. The ages are 45, 42, 14, and 9. You try this code:

```
int totalAges = 45 + 42 + 14 + 9;
double average = totalAges / 4;
System.out.println( "The average age is " + average );
```

The output is 27.0. When you check the result using a calculator, you find that the correct average is 27.5. What is wrong with this code? Record your response in the space provided.

_____
_____
_____
_____
_____
_____
_____
_____
_____

Lab 5-3
Output Results

In this lab, you will refine the appearance of output by applying the String.format( ) method. Before beginning this activity, download the files for this lab from the student companion website.

Learning Goals
- Format output for a clear presentation of results.
- Design a test plan.

Materials
- jGRASP Integrated Development Environment
- Starter files from the student companion website

Application and Extension of Knowledge

Calculating Total Price

You are buying a gift for a birthday party. You want to calculate the total price of the gift including sales tax to be sure you have enough cash to pay for it. Edit the GiftPriceSnippet.java program to input the price of the gift and the sales tax rate. For example, if the sales tax is 5 percent, the user will enter 5. Output the original and final prices, formatted as money, and output the sales tax rate formatted as a percentage.

Procedure

1. Follow this algorithm, and plan your solution in the space provided. Write generalized statements that will help you design the code later.

   ```
   /***** 1. Input the price of the gift and the sales tax rate. */
   ```

   ___

   ___

   ```
   /***** 2. Convert the sales tax to a decimal, calculate the sales tax, and
               add that to the gift price.
   */
   ```

   ___

   ___

   ```
   /***** 3. Output the original and final prices formatted as money.
               Output the sales tax rate as a percentage.
   */
   ```

   ___

   ___

2. Launch jGRASP, and open GiftPriceSnippet.java starter file.

3. Enter code for the generalized statements you wrote above.
4. Test the program, and correct any errors as needed.

Reflections

1. Compare and contrast the benefit of using String.format( ) with the extra effort required to set up the code.

___
___
___

2. How would you format the output so the prices align if several gifts were purchased at the same time?

___
___
___

Creating an Invoice

You have a lawn-care business and need to invoice your client for a month's worth of work. You charge $25 each time you mow the lawn and $10.50 for each weeding. Edit the LawnCareSnippet.java program to input the number of hours you worked in each activity. Create an invoice where the output looks like the sample below.

```
<Your Name> Lawn Care
Activity     Number      Fee($)      Total ($)
Mowing       5           25.00       125.00
Weeding      3           10.50       31.50

Total due : $156.50
```

Procedure

1. Follow this algorithm, and plan your solution in the space provided. Write generalized statements that will help you design the code later.

```
/***** 1. Define the fees for mowing and for weeding as constants. */
```

___
___
___

```
/***** 2. Input the number of times of mowing and weeding. */
```

___
___
___

```
/***** 3. Calculate the fees for mowing and weeding and total due. */
```

___
___
___

```
/***** 4. Output the formatted results. */
```

_____

_____

_____

2. Launch jGRASP, and open LawnCareSnippet.java starter file.
3. Enter code for the generalized statements you wrote above.
4. Test the program, and correct any errors as needed.

Reflections

1. If you add another activity, how would you need to change the formatting of the charge for that fee?

   _____

   _____

   _____

2. Compare the advantages and disadvantages of outputting the header line as one string containing spaces between headers versus formatting each header as a separate string.

   _____

   _____

   _____

Calculating Volume and Surface Area of a Sphere

Given the radius of a sphere, the volume and surface area of the sphere can be calculated. The formulas are as follows.

volume = $4/3\pi r^3$

surface area = $4\pi r^2$

Using the formulas above, calculate the volume and surface area of the sphere and output the results to three decimal places. Edit the SphereSnippet.java program to input the radius. Additionally, prepare a test plan to use to verify your output.

Procedure

1. Follow this algorithm, and plan your solution in the space provided. Write generalized statements that will help you design the code later.

   ```
   /***** 1. Define pi as a constant: 3.14159265 */
   ```

   _____

   _____

   _____

   ```
   /***** 2. Input the radius of the sphere */
   ```

   _____

   _____

   _____

```
/***** 3. Calculate the volume and surface area of the sphere. */
```

_____

_____

_____

```
/***** 4. Output the results formatted to three decimal places. */
```

_____

_____

_____

2. Launch jGRASP, and open **SphereSnippet.java** starter file.
3. Enter code for the generalized statements you wrote above.
4. Test the program, and correct any errors as needed.

Reflections

1. How many places of pi do you think are required for this exercise? Explain your answer.

_____

_____

_____

2. How well did the test plan predict the output? Explain your assessment.

_____

_____

_____

Debugging Challenge

You want to format a discount value as a percentage. To do so, you try this code:

```
double discount = .20;
String s = String.format( "Today's discount is %.0f%", discount * 100 );
System.out.println( s );
```

When this code runs, it generates an exception. The message is:

```
UnknownFormatConversionException: Conversion = '%'
```

What is wrong with the code? Record your response in the space provided.

_____

_____

_____

_____

_____

_____

Name _____ Date _____ Class _____

CHAPTER 6
Classes

Classes may be called object descriptors. They are a framework from which objects can be created. The class describes the data each object will have. The class also provides methods (code) to store, get, and manipulate the data in the object. Keeping the data and methods together in a class is called encapsulation. The root of this word is *capsule*. Encapsulation isolates the data in a "capsule" and keeps it safe from intrusions. This lab will help you to use classes to instantiate objects and apply the class methods to that object.

Chapter Highlights

- An advantage to using classes is some programmers can write a class that many programmers can reuse.
- The *application programming interface (API)* tells programmers wishing to use the class how to create objects of that class, what methods are available, and how to call those methods.
- The function of the *constructor* is to create the object and make sure the object has valid values for all its data.
- The *object reference* is the name of the object and holds a location in memory that is used to find the object.
- The Turtle class allows the programmer to create a turtle sprite, or image, that can move around a window to draw shapes. This is done by calling methods to move the turtle and change the turtle's direction.
- To use the Turtle class, there needs to be at least four .java files in the same folder: Turtle.java, Sprite.java SpriteAnimator.java and another Java file that is the application you write to direct the turtle's drawing.
- The Java Class Library has a rich set of predefined classes that can be used to create special features in your programs.

(continued)

While studying, look for the activity icon 📲 for:
- Vocabulary terms with e-flash cards and matching activities.
- Starter files for lab activities.

These activities can be accessed at
www.g-wlearning.com/informationtechnology/1773

(continued)
- A ***package*** is a collection of related classes. To use many of the classes in the Java Class Library, it is necessary to import the class.
- Scanner, String, Random, Math, and Color are examples of Java classes.

Warm-Up Exercises

Questions 1–5 refer to the following information.
Consider the following API for encapsulating and managing information about a virtual box.

Class
```
Box
```

Constructor
```
Box( int length, int width, int height )
```

Methods
```
void setLength( int length )
int getLength( )
void setWidth( int width )
int getWidth( )
void setHeight( int height )
int getHeight( )
int surfaceArea( )
int volume( )
```

_____ 1. What is the proper statement to instantiate an object named box?
   A. Box box = new Box(5, 8, 9);
   B. box = Box(5, 8, 9);
   C. Box box = new Box( );
   D. box = new Box(5, 8, 9);
   E. Box box = new Box(length=5, width=8, height=9);

_____ 2. What is the proper statement to calculate the surface area of an instance named box?
   A. int area = box.surfaceArea(l, w, h);
   B. int area = box.surfaceArea( );
   C. int area = surfaceArea(box);
   D. int area = new box.surfaceArea( );
   E. int area = getBox.surfaceArea( );

_____ 3. What is the proper statement to update the length of an instance named box?
   A. int length = box.setLength( );
   B. box.getLength(14);
   C. box.setLength(14);
   D. int length = new box.Length( );
   E. int l = box.getLength( );

_____ 4. What is the proper method call to calculate the volume of an instance named box?
   A. int volume = length * width * height;
   B. box.getVolume( );
   C. box.setVolume(length * width * height);
   D. box.volume( );
   E. getVolume( );

_____ 5. What is the output from the following statements?
```
Box bigBox = new Box( 10, 10, 10 );
System.out.println( "Volume = " + bigBox.volume( ) );
```
   A. Volume = bigBox(1000)
   B. box.Volume = 1000
   C. Volume = 1000
   D. Volume = 1000.0
   E. Volume = 1,000

Questions 6–10 refer to the following information.

Few students want to be the first to make their presentation. Professor Hart chooses the order for student presentations using his program called the Randomalizer. Consider the following code segment that uses the Random class.

```
Random rand = new Random( );
int presenter = rand.nextInt( 22 );
```

_____ 6. How many students can the Randomalizer accommodate with the above code?
   A. 22
   B. 23
   C. 21
   D. 11
   E. 0

_____ 7. What is the first number generated?
   A. 22
   B. 0
   C. 21
   D. 11
   E. Impossible to determine.

_____ 8. What is the lowest number generated?
   A. 22
   B. 23
   C. 1
   D. 0
   E. Impossible to determine.

_____ 9. What is the highest number generated?
   A. 21
   B. 23
   C. 1
   D. 0
   E. Impossible to determine.

_____ 10. What is the most frequently generated number?
   A. 22
   B. 23
   C. 1
   D. 0
   E. Impossible to determine.

Lab 6-1

Introduction to Classes

The mainstay of object-oriented programming is the class. The class encapsulates data and methods into a packet of code called an object. The key skills for working with classes are using constructors, passing arguments, calling methods, and handling return values. The Turtle class can help in exercising these important skills. Before beginning this activity, download the files and folders for this lab from the student companion website.

Learning Goals
- Use an API to add an object and its methods to a program.
- Inspect and study errors to avoid them in the future.
- Apply new methods from the Turtle class.

Materials
- jGRASP Integrated Development Environment
- Starter files from the student companion website
- Turtle API (instructor-supplied handout)

Application and Extension of Knowledge

Using a Game Object Class with a Client Class

In a multiplayer game, there is a need to keep each player's statistics separate from the other players, but all players have common needs. To keep it simple, a Player class is created. Each player is represented by a different object. In a certain game program, the code needs to keep track of the player name, which is unique to each player; the level reached; the inventory collected; and the merit awards earned.

The class client, DestinyAscentSnippet.java, is a sandbox, or practice area, for exercising the facilities of the Player class. Use the following API for the Player class. Be sure the Player.java and DestinyAscentSnippet.java files from the student companion website are both located in your working folder.

Player Constructor	Action
Player(String username)	Creates a Player object with username, a level: 1, inventory: "rope" and merit awards: "Beginner".

Method	Action
String getUsername( )	Returns player's name.
int getLevel( )	Returns player's level in game.
void bumpLevel( )	Increments the player's level by 1.
String getInventory( )	Returns a String containing player's inventory items.
void addToInventory(String newItem)	Adds a new item to the player's inventory.
String getMerits( )	Returns a String containing player's merit awards.
void addToMerits(String newAward)	Adds a new item to the player's merit awards.
String toString( )	Returns a formatted String for displaying current values of the Player object data.

Procedure

1. Launch jGRASP, and open the Player.java file.
2. Compile the file. The Player.class file is generated. This is the file used for the API.
3. Open the DestinyAscentSnippet.java file. Notice that the only class imported is the Scanner class. The compiler will look for Player.class in the same folder as the client file. You only need to import classes you use from the Java Class Library.
4. At comment 1, write the statement that creates an instance of the Scanner class. Name the object input.
5. At comment 2, get a username from the player. First, prompt the player for what the program wants.
6. At comment 3, use the Player constructor to define the player1 object with the name captured by the code written for comment 2. Refer to the Player API for the syntax.
7. At comment 4, write a statement that prints the instance values for player1. Refer to the API for the toString( ) method.
8. Compile and run the file. In the space that follows, note any errors in coding that occurred. This will help you to remember not to make them again.

   _____

   _____

   _____

9. During the actual Destiny Ascent game, players will advance to the next level. Use the API for bumpLevel( ). Apply what you have learned about the use of an object's method. At comment 5, write the object name, a dot, and the method. Then, print out the current level.
10. Compile and run the file. Make any notable comments about errors in the space that follows.

    _____

    _____

    _____

11. At comment 6, apply what you have learned to add an inventory item. The API shows the addToInventory( ) method takes a String argument of the new item. Use "ID tag" for the new item. Add white space where appropriate. Compile and run the file.
12. At comment 7, apply what you have learned to add a new merit award. The API shows the addToMerits( ) method takes a String argument of the new award. Use "Novice Climber" for the new award. Add white space where appropriate. Compile and run the file.
13. At comment 8, use the toString( ) method again to show the current player status. Add white space where appropriate. Compile and run the file. You have used a client file to apply the methods in a class file. Make notes in the space that follows about what you learned.

    _____

    _____

    _____

    _____

    _____

Reflections

1. How do you think using a Player class might simplify the code for a multiplayer game?

   _____
   _____
   _____

2. Demonstrate how you would create an object and call a method for player2.

   _____
   _____
   _____

Exploring Turtle Graphics

In this chapter, the textbook activities had you creating turtle graphics using several of the methods from the Turtle class API. You can be very creative using those methods and trying out other methods, even instantiating several turtle objects. In this lab, you will practice using some new methods in the Turtle class. Then, you will write the code to generate your own turtle graphics.

Procedure

1. Launch jGRASP, and open the SpriteAnimator.java, Sprite.java, and Turtle.java files located in the TurtleHijinks folder. Compile them in this order:

   A. SpriteAnimator.java
   B. Sprite.java
   C. Turtle.java

   You may then close these three .java files.

2. Open the TurtleHijinks.java file in the TurtleHijinks folder. Compile and run the file. In the space below, describe what you see in the graphics window.

   _____
   _____
   _____

3. Close the graphics window, and locate comment 1 in the buildScript( ) method in the TurtleHijinks.java file. Apply what you have learned to instantiate a Turtle object named shelly at the location (350, 250) in the graphics window. Compile and run the file. Report what you see.

   _____
   _____
   _____

4. Locate comment 2 in the buildScript( ) method. Apply what you have learned to define a double constant named DISTANCE with a value of 90. What is the keyword that defines a constant in Java?

   _____

Name _____

5. Locate comment 3 in the buildScript( ) method. Apply what you have learned to make the shelly object draw a triangle by turning left 120 degrees and drawing a line DISTANCE steps long three times. Click the red runner icon to recompile and run the application. On a piece of paper, make a sketch of what you see. Describe it below.

_____

_____

6. Locate comment 4 in the buildScript( ) method. Apply what you have learned to instantiate a second Turtle object name crawly at the location (450, 250) in the graphics window. Then, recompile and run the application. What color is the new turtle? What color do you predict the path will be?

_____

_____

7. Locate comment 5 in the buildScript( ) method. You can change the drawing color. Examine the setDrawingColor(Color color) method in the Turtle class API. To find available colors, search the Internet for oracle docs javafx color constants. To use a color constant for the argument Color color, replace the variable color with a color constant in the table. For example, for orange, use Color.ORANGE. Write the statement to change the crawly object's drawing color to ORANGE. In the space below, record the URL for the Oracle web page where you found the color constants.

_____

_____

8. In the Turtle class API, there is a setSpeed( ) method to speed up or slow down the animation of the turtle. Locate that method, and note the choices for arguments for the method. Write the static constants in the space below.

_____

_____

_____

9. Locate comment 6 in the buildScript( ) method. The default drawing speed is Turtle.MEDIUM. Set the drawing speed of the crawly object to Turtle.FAST. What is this statement?

_____

_____

10. Locate comment 7 in the buildScript( ) method. Apply what you have learned to draw a hexagon using the crawly object. At each vertex, turn left 60 degrees and go forward DISTANCE steps. Recompile and run. On a piece of paper, make a sketch of what you see. Describe it below.

_____

_____

11. Now, add a little story to this application. Shelly is so impressed with Crawly's hexagon that she moves over to Crawly's drawing and erases the triangle. To code this story element, locate comment 8 in the buildScript( ) method. Follow the steps in the comment using methods from the Turtle class API. Recompile and run the application. On a piece of paper, make a sketch of what you see. Describe it below.

_____
_____
_____

Reflections

1. Why do you think additional turtles are colored differently?

_____
_____
_____

2. Explain how you could get both turtles to face in the same direction at the end of this application. Describe a method you would add to the Turtle class API that would set a turtle's heading to a value of degrees.

_____
_____
_____

Making an Original Turtle Drawing

Apply what you have learned so far about Java and the Turtle class to create an idea for a turtle animation. Write an algorithm to code this animation. Use the space below to brainstorm and write the algorithm. Use the starter file TurtleAnimation.java to create the application. Write ample comments in your code to remind you and inform anyone reading the code what the steps are doing. Supply the four Java files required for this animation for grading.

_____
_____
_____
_____
_____
_____
_____
_____
_____
_____

Debugging Challenge

Consider the following buildScript( ) method code. It does not compile cleanly. Find the errors. Suggest a fix in the space below.

```
35 public void buildScript( ) {
36
37     // create 3 Turtle objects
38     Turtle bertle = new Turtle( root, 200, 200 );
39     Turtle certle = new Turtle( root, 250, 200 );
40     Turtle dertle = new Turtle( root, 300, 200 );
41
42     // change the drawing color of one turtle.
43     bertle.setDrawingColor( CADETBLUE );
44
45     // move all 3 turtles
46     bertle.forward( 10 );
47     certle.backwards( 10 );
48     dertle.moveTo( bertle.getX( ), certle.getY( ) );
49
50 } // end buildScript
```

Lab 6-2

Java Class Library

In this lab, you will apply the rules for arithmetic operations. You will solve problems requiring expressions. Before beginning this activity, download the files for this lab from the student companion website.

Learning Goals

- Research and incorporate methods from the Java Class Library.
- Describe the structure of a page from the Java Class Library.

Materials

- jGRASP Integrated Development Environment
- Starter files from the student companion website

Application and Extension of Knowledge

Using the Random Class

The Random class can generate seemingly random numbers. In the chapter activities in the textbook, you created new random-number generators with the Random class. Then, you used those objects to generate integers within a specified range. In this activity, you will explore other methods available in the Random class.

Procedure

1. Search the Internet for oracle docs java random. Visit the Oracle Docs web page for the Random class. Respond to the prompts in the spaces that follow with information from that page.

    A. Package that Class Random belongs to:
    _____

    B. Number of constructors:
    _____

    C. What does the Constructor Detail tell you about a seed?
    _____
    _____
    _____
    _____

2. Launch jGRASP, and open the LotsOfRandom.java file. Locate comment 1, and follow the directions.

3. Compile and run the program. Record the five integers in the space below. Run the program again. Record those five integers as well. What do you observe?
    _____
    _____
    _____
    _____

4. Edit the constructor to include the long seed defined on line 11 in the program. Recompile and run the program. Record the five integers in the space below. Run the program again. Record those five integers as well. Check with your classmates to see what their outputs are. What do you observe?
    _____
    _____
    _____
    _____
    _____

5. Look through the Method Detail for the Random class. Locate and use the methods to generate and output a random byte (not bytes), random long, random boolean, random float, and random double. Prepare meaningful output statements. Recompile and run the file. Show your output below.

___

Reflections

1. Explain why you think someone would want to generate the same random numbers time after time. Do you think doing so defeats the notion of randomness?

___

2. Look up the Math.random( ) method in the Math class API in the Java Class Library. Describe the differences between that and the Random class.

___

Using the Math Class

Perhaps the most useful attribute of the Math class is that the methods and constants are static. The programmer does not need to define a Math object to use them. For example, to get a random number using the Math class method, simply call the method and assign the result to a double variable. However, the Math.random( ) method returns a number between 0 and 1. Some manipulation will be required to yield the values desired. Consider using Math.random( ) to roll dice. The result should be an integer from 1 to 6. In this activity, you will explore how to accomplish this.

Procedure

1. Search the Internet for oracle docs java random. Visit the web page for the Math class API. In the spaces that follow, write the API for the designated methods with information from the Method Summary on that page. For more information about any of these methods, click the method name to go to the information page for that method.

   A. max( )

   ___

B. min( )

C. random( )

2. Launch jGRASP, and open the LotsOfMath.java file. Locate comment 1, and follow the directions. Compile and run the file. In the space that follows, explain how you determined which methods to use.

3. So you do not have to keep providing input, delete the end of comment mark at the end of comment 1. In the space that follows, explain what this does to the code.

4. Locate comment 2, and follow the instructions to generate a random number. If the goal is a die roll, how helpful is this result? Comment on the output in the space that follows.

5. Locate comment 3. In order to take the random double and produce a number in a range, first multiply the quantity of numbers in the range and add the lowest number in the range. For example, to generate random numbers from 1 to 6, use Math.random( ) * 6 + 1. There are six numbers in the range, and you want to start at 1. Write the code to generate random numbers between 1 and 6. Comment on the output and its usefulness for a die roll in the space that follows.

6. Locate comment 4. Apply what you learned about casting and follow the instructions. Comment in the space that follows on the worthiness of this output for use as a dice roll.

Name _____ Chapter 6 Classes 77

Reflections

1. Compare the ease of use of the Random class with that of Math.random( ).

   _____
   _____
   _____

2. Describe the scenarios for using the various overloaded methods for min( ) and max( ).

   _____
   _____
   _____

Debugging Challenge

Consider the following program. It does not compile.

```java
/* Debugging Challenge Lab 6-2
   Your name here
*/

import java.util.Scanner;
public class DebuggingChallenge6_2 {
   public static void main( String [ ] args ){

      /***** 1. Get two double numbers from the user
                and determine the maximum and the minimum
                using the Math.max( ) and Math.min( ) methods.
                Output the results.   */
      Scanner input = new Scanner( System.in );
      System.out.print( "Enter first double value: " );
      double double1 = input.nextdouble( );
      System.out.print( "Enter second double value: " );
      double double2 = input.nextdouble( );

      System.out.println( "The larger of the two values is " +
         Math.max( double1, double2 ) );
      System.out.println( "The smaller of the two values is " +
         Math.min( double1, double2 ) );

   } // end main
} // end class
```

The compiler displays these errors:

```
DebuggingChallenge6_2.java:18: error: cannot find symbol
      double double1 = input.nextdouble( );
                            ^
  symbol:   method nextdouble()
  location: variable input of type Scanner
DebuggingChallenge6_2.java:20: error: cannot find symbol
      double double2 = input.nextdouble( );
                            ^
  symbol:   method nextdouble()
  location: variable input of type Scanner
errors
```

What is wrong with this code? Record your responses in the space provided.

Name _____ Date _____ Class _____

CHAPTER 7
Drawing

A very popular use of Java is for app development. By incorporating images, formatted text, and shapes, JavaFX brings graphical user interfaces to the user. Interactions are discussed in a later chapter. This chapter relates to incorporating text, color, and shapes into an application.

Chapter Highlights

- The three steps to creating a JavaFX app are: create the stage, set the scene, display the stage.
- The Application class provides methods that start the application and manage the graphics you create.
- A group is a virtual container within the scene that holds everything in the JavaFX application.
- A scene controls the layout of all items in all groups.
- A stage is the window created by the start method where the graphical output is displayed.
- The scene graph organizes all the effects in a scene. It is an internal tree of nodes and groups.
- The JavaFX Text and Font classes are used to work with text, including selecting the typeface.
- Millions of colors can be generated by mixing a combination of red, green, and blue (RGB).
- The *fill* is the color on the inside of a character or shape and the *stroke* is the outline of a character or shape.
- Vector graphics are composed of shapes based on mathematical equations.
- Irregular shapes can be drawn using the JavaFX Path class.

While studying, look for the activity icon 📱 for:
- Vocabulary terms with e-flash cards and matching activities.
- Starter files for lab activities.

These activities can be accessed at
www.g-wlearning.com/informationtechnology/1773

Warm-Up Exercises

_____ 1. Consider the following code segment. What color is displayed as a result of executing this code segment?

```
Group root = new Group( );
Color sceneColor = Color.rgb( 0, 255, 0 );
Scene s = new Scene( root, 300, 300, sceneColor );
```

A. RED
B. ORANGE
C. YELLOW
D. GREEN
E. BLUE

Questions 2 through 4 refer to the following incomplete class definition.

```
public class SampleApp extends Application {
   @Override
   public void start( Stage mainStage ) {
      Group root = new Group( );
      Scene scene = new Scene( root, 300, 200 );
      mainStage.setTitle( "App 1" );
      mainStage.setScene( scene );
      mainStage.show( );

   } // end start method
} // end Class
```

_____ 2. What is the name of the application?

A. App 1
B. mainStage
C. SampleApp
D. scene
E. root

_____ 3. What is the width of the application window?

A. 600 pixels
B. 200 pixels
C. 300 bytes
D. 200 bytes
E. 300 pixels

_____ 4. What is the title of the application window?

A. App 1
B. mainstage
C. SampleApp
D. scene
E. root

_____ 5. Which color constant is equivalent to the following code?
```
Color.rgb( 0, 0, 0 );
```
   A. Color.RED
   B. Color.GREEN
   C. Color.BLUE
   D. Color.BLACK
   E. Color.WHITE

_____ 6. With which JavaFX class are the MoveTo( ) and LineTo( ) classes used?
   A. Rectangle
   B. Draw
   C. Graph
   D. Path
   E. Ellipse

_____ 7. Which JavaFX shape class can draw squares on the stage?
   A. Rectangle
   B. Draw
   C. Graph
   D. Arc
   E. Ellipse

_____ 8. Of what is an arc a section?
   A. Rectangle
   B. Draw
   C. Graph
   D. Path
   E. Ellipse

_____ 9. What property does the Circle class have that describes the size of the circle?
   A. stroke
   B. center
   C. fill
   D. radius
   E. shape

_____ 10. Which JavaFX shape class has the methods setStartX( ) and setEndY( )?
   A. Arc
   B. Line
   C. Rectangle
   D. Circle
   E. Triangle

Lab 7-1

Java Graphics Components

The basic component of a JavaFX graphics application is the application window. In this chapter, you learned how to change the title and display the window. Using the Java Class Library, you will identify other properties and methods that apply to the window. In this lab, you will extend your knowledge of the JavaFX application window. Before beginning this activity, download the files for this lab from the student companion website.

Learning Goals
- Manage the size of the JavaFX application window.
- Set the style of the JavaFX application window.

Materials
- jGRASP Integrated Development Environment
- Starter files from the student companion website

Application and Extension of Knowledge

Resizing the JavaFX Application Window

Defining the application window in JavaFX depends on the target platform. If you are designing for a mobile app, you need to design for landscape and portrait orientations. If you are designing for Windows or an iMac, you need to plan for a user resizing or scaling a window and its contents. If you are designing for a tablet, similar considerations must be judged. Not only that, but different platforms also have different screen sizes. In this lab, you will adjust scene sizes and the titles.

Procedure

1. Launch jGRASP, and open the JavaFXAppWindowSnippet.java file. Locate comment 1. Follow the directions. Then, compile and run the application. Describe the results.

   _____

   _____

   _____

2. Locate comment 2. In order to make a new scene definition, you must first comment out the old one. Delete the */ at the end of comment 1. The comment now extends to the end of comment 2. Enter the new scene definition as specified in comment 2. Do not forget to update the stage title. Then, recompile and run the application. Describe the results in the space provided.

   _____

   _____

   _____

Reflections

1. Why do you think an app developer would decide to support only a few platforms for a specific app?

   _____
   _____
   _____

2. Why do you think an app for a PC or Mac may *not* need to start out full-screen?

   _____
   _____
   _____

Setting the Stage Style

Have you noticed that some pop-up windows do not have a title bar, and accordingly do not have the minimize, restore, and exit buttons? These windows have internal coding to permit closing the app. The StageStyle property of the stage provides options for the decoration of a window. Commonly, "decorations" are the title bar and the buttons for manipulating the window. In each platform, Windows, Mac, and Linux, they look just a bit different. The API from Oracle Docs Java Class Library is below.

- StageStyle.DECORATED: a stage with a white background and platform decorations; default.
- StageStyle.UNDECORATED: a stage with a white background and no decorations.
- StageStyle.TRANSPARENT: a stage with a transparent background and no decorations.
- StageStyle.UTILITY: a stage with a white background and minimal platform decorations.

The initStage( ) method takes these constants as arguments. The stage style must be set before the show( ) method is called. In this lab, you will experiment with StageStyle settings.

Procedure

1. Launch jGRASP, and open the JavaFXStageStyleSnippet.java file. Locate comment 1, and enter the following code.

   ```
   mainStage.initStyle( StageStyle.UNDECORATED );
   ```

   Compile and run the application. Describe the results in the space that follows.

   _____
   _____
   _____

2. A problem arises! There is no title bar and, therefore, no exit button to close the window. Click the **End** button on the **Run I/O** tab in jGRASP to terminate the app.

3. Change the StageStyle property to TRANSPARENT. Recompile and run the application. Describe the results in the space that follows.

   _____
   _____
   _____

4. Change the **StageStyle** property to **UTILITY**. Recompile and run the application. Describe the results in the space that follows.

5. One of the properties of the stage allows a **boolean** to control whether or not the window can be resized. Set the **StageStyle** property to **DECORATED**. Then, locate comment 2, and write a statement to set the mainStage.setResizable property to false.

   ```
   mainStage.setResizable( false );
   ```

   Recompile and run the application. Describe the results in the space that follows.

Reflections

1. Why do you think a developer would select one of the stage styles with no buttons or disabled buttons?

2. Can you find another way to close an **UNDECORATED** window?

Debugging Challenge

Consider the following erroneous code.

```
/*  JavaFX Debugging Challenge - Lab 7-1
    your name here
*/

import javafx.application.Application;
import javafx.scene.Group;
import javafx.scene.Scene;
import javafx.stage.Stage;

public class DebuggingChallenge7_1 extends Application {

  @Override
  public void start( Stage mainStage ) {
    Group root = new Group( );
    Scene scene = new Scene( root, 400, 200 );

    mainStage.setTitle( "JavaFX Window" );
    mainStage.setScene( stage );
    mainStage.show( );
  } // end start method

  public static void main( String [ ] args ) {

    launch( args );

  } // end main
} // end class
```

When compiled, the following error message is displayed.

```
DebuggingChallenge7_1.java:18: error: cannot find symbol
    mainStage.setScene( stage );
                        ^
    symbol: variable stage
    location: class DebuggingChallenge7_1
1 error
```

Explain how to correct the error in the space that follows.

_____

_____

_____

_____

_____

Lab 7-2

Text and Color

In this chapter, you learned a number of techniques to add text to a JavaFX app. By examining the Java Class Library, you will be able to incorporate many more features into your apps. The ability to read the API is a skill that will keep you learning throughout your career as a computer programmer. To be a great programmer, you need to be a lifelong learner. Before beginning this activity, download the files for this lab from the student companion website.

Learning Goals
- Identify new techniques using the Text, Paint, and Font classes.
- Incorporate new techniques using the Color class.

Materials
- jGRASP Integrated Development Environment
- Starter files from the student companion website

Application and Extension of Knowledge

Setting Properties for Text

On occasion, you will find you want to adjust the space between lines for text in a window. In this lab, you will experiment with the Text class setLineSpacing( ) method.

Procedure

1. Search the Internet for oracle docs javafx text. View the web page for the Text class. Take care not to select the page for the TextField class. Record the URL for the Text class API in the Java Class Library. Then, list two properties and two methods found on the page.

   _____
   _____
   _____

2. Launch jGRASP, and open the JavaFXTextPropertySnippet.java file. Examine the code provided. Then, compile and run the application. Describe the results in the space that follows.

   _____
   _____
   _____

3. Locate comment 1. Enter a statement to set the line spacing to 5. Recompile and run the application. Experiment with other values. Describe the results in the space that follows.

   _____
   _____
   _____

4. Identify another method that sets a property. Implement that method. Describe what you did and the results in the space provided.

   _____
   _____
   _____
   _____
   _____
   _____

Reflections

1. Consider why there are so many different properties defined. Which ones might you use in the future? Can you think of any characteristics of text that are not supported?

   _____
   _____
   _____

2. How is the Oracle Docs web page for JavaFX Text similar or dissimilar to other pages in the Java Class Library? Do you think this is a good thing?

   _____
   _____
   _____

Adjusting Color Brightness

When a color is set for a JavaFX object, it may be displayed too light or too dark. Rather than going back to adjust the RGB color contributions, you can use the darker( ) or lighter( ) method. This lab experiments with these two methods.

Procedure

1. Search the Internet for oracle docs javafx color. View the web page for the Color class. Record the URL for the Color class API. Then, list two color constants, two properties, and two methods found on the page.

   _____
   _____
   _____

2. Launch jGRASP, and open the JavaFXColorSnippet.java file. Locate comment 1, and follow the directions. Do not forget to add the rectangles to the effects list. Then, compile and run the application. Describe the results in the space that follows.

   _____
   _____
   _____

3. Color.CADETBLUE is a Color object defined as a constant in the Color class. You can apply the darker( ) method to it as you would use any object reference to call a method:

   ```
   Color.CADETBLUE.darker( )
   ```

   Locate comment 2 and follow the directions. Enter a statement to set the fill of rect2 to Color.CADETBLUE.darker( ). Then, recompile and run the application. Describe the results in the space that follows.

   _____
   _____
   _____

4. Locate comment 3 and follow the directions. Enter a statement to set the fill of rect2 to a lighter shade of CADETBLUE. Then, recompile and run the application. Describe the results in the space that follows.

___

___

___

5. Locate comment 4 and follow the directions. Define **Text** objects to label the rectangles. Do not forget to add them to the effects list. Then, recompile and run the application. Describe the results in the space that follows.

___

___

___

Reflections

1. Consider the pairs of color constants GRAY and GREY, DARKGRAY and DARKGREY, and LIGHTGRAY and LIGHTGREY. Looking at the API, examine the color contribution for each pair of constants. Describe how they are related. Explain the reasoning for this duplication.

___

___

___

2. Explain the meaning of the sections on the **Text** class page in the Java Class Library that are titled "Methods inherited from class javafx.scene.shape.Shape," "Methods inherited from class javafx.scene.Node," and "Methods inherited from class java.lang.Object."

___

___

___

Debugging Challenge

Consider the following faulty code segment.

```
/***** 1. Draw three Rectangles named rect1, rect2, and rect3.
         Set the fill color of rect1 to CADETBLUE */
Rectangle rect1 = new Rectangle( 20, 20, 200, 100 );
rect1.setFill( Color.CADET BLUE );
Rectangle rect2 = new Rectangle( 20, 130, 200, 100 );
Rectangle rect3 = new Rectangle( 20, 240, 200, 100 );
```

When compiled, these errors are reported:

```
JavaFXColorAnswer.java:25: error: ')' expected
     rect1.setFill( Color.CADET BLUE );
                                ^
JavaFXColorAnswer.java:25: error: not a statement
     rect1.setFill( Color.CADET BLUE );
                                ^
JavaFXColorAnswer.java:25: error: ';' expected
     rect1.setFill( Color.CADET BLUE );
                                ^
3 errors
```

Name _____   Chapter 7 Drawing   **89**

How many errors need to be corrected? How can the code be fixed?

_____

_____

_____

_____

_____

Lab 7-3

JavaFX Shapes

The basic shapes in JavaFX can be combined in innovative ways to build icons and other symbols for business or games. Add color and text, and the effects can be stunning. It takes a great deal of planning and computational thinking in advance of coding the app. In this lab, preliminary drawings and algorithm development make the task of coding a breeze. Before beginning this activity, download the files for this lab from the student companion website.

Learning Goals
- Create a custom button using JavaFX shapes.
- Design a playing piece for a classic game in JavaFX.

Materials
- jGRASP Integrated Development Environment
- Starter files from the student companion website
- Grid paper (instructor-supplied handout)

Application and Extension of Knowledge

Designing a Dominoes Game Piece

The game dominoes has its origins in ancient China. It uses tiles with circles to represent two numbers between 1 and 6. The figure shows the tile for a domino with pips (dots) for a two and a five (2/5). In this lab, you will build this prototype of a domino tile. You will engage in a thought exercise to think about how to build an entire set of dominos.

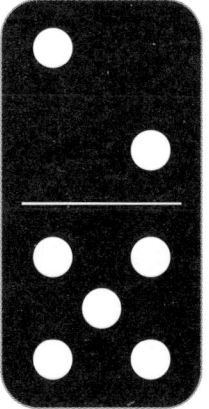

Goodheart-Willcox Publisher

Procedure

1. Study the image of the domino. What symmetries are involved? How can you make the spacing even? Record your thoughts in the space below.

   _____

   _____

   _____

2. The sketch below will direct the algorithm for drawing the 2/5 domino. Write the coordinates relative to the rectangle for the centers of the circles in the space provided. The first one is provided for you. The identifier pip2_1 indicates it is the pip from the 2 side of the tile and it is the first pip. The next pip is pip2_2. For the 5 side, the identifiers are pip5_1 and pip5_2 on the top row, pip5_3 and pip5_4 on the bottom row, and pip5_5 in the middle.

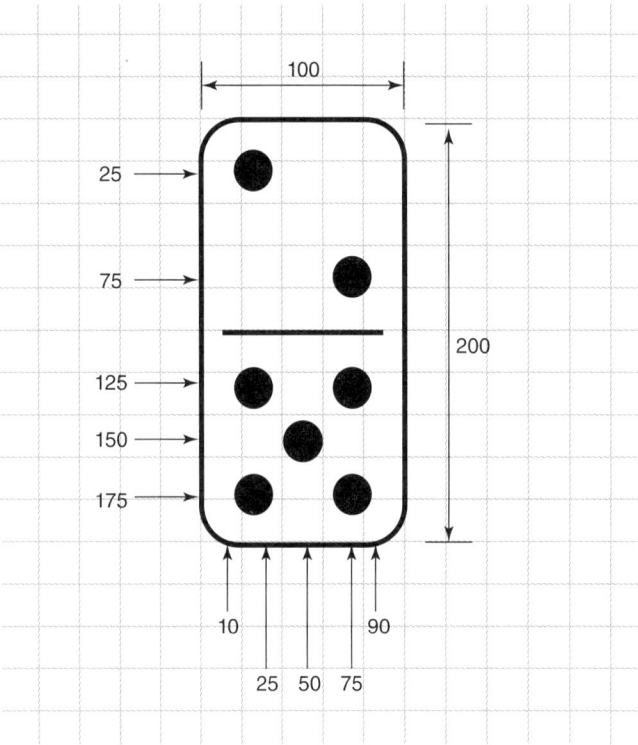

Goodheart-Willcox Publisher

pip2_1: center (25, 25), radius 10

_____

_____

_____

_____

_____

_____

_____

3. Launch jGRASP, and open the JavaFXDominoesSnippet.java file. Locate comment 1, and follow the directions. Compile and run the application. Describe the results in the space that follows.

    _____
    _____
    _____

4. Locate comment 2, and follow the directions. A line object is drawn using the coordinates of the endpoints as arguments. The line is horizontal, so the two *y* coordinates are the same. Define a double variable lineY, and set that to be the *y* coordinate of the domino plus half the height of the domino. The *x* coordinates are 10 pixels in from each edge. Use these coordinates to instantiate a Line object, set the strokeWidth to 3, and set the Stroke to Color.GRAY. Add the Line object to the list of effects. Compile and run the application. Describe the results in the space that follows.

    _____
    _____
    _____

5. Locate comment 3, and follow the directions. It is important to avoid magic numbers, so define constants for each of the *x* and *y* coordinates of the circles. The radius and *x* coordinates have been defined for you.

6. Locate comment 4, and use the constants to define the circles for the 2 side of the tile. The coordinates of the rectangle and the first circle have been defined for you. Add the circles to the effects list. Then, compile and run the application. Describe the results in the space that follows.

    _____
    _____
    _____

7. Locate comment 5, and use the constants to define the circles for the 5 side of the tile. Add the circles to the effects list. Then, compile and run the application. Describe the results in the space that follows.

    _____
    _____
    _____

Reflections

1. In gameplay, a domino can be rotated. Explain how that might affect your code.

    _____
    _____
    _____

2. Explain why so many constants were defined for this JavaFX drawing.

    _____
    _____
    _____

Making a Custom Button for an App

Think about apps you use. Do they have standard-looking buttons, or have the developers designed buttons that evoke the product or game being used? Chances are each button has a unique look. In this lab, you will develop a logo and design a button with text on it, as shown. The button will be the graphic only. It will not be interactive. You will learn how to make it interactive in later chapters.

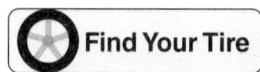

Find Your Tire

Goodheart-Willcox Publisher

Procedure

1. Think of a product or game with which you are familiar, or make one up. Design a logo that you can build using JavaFX Shapes. The button base should be a rounded rectangle. Make the sketch on grid paper. Write your name on the grid paper. Here is an example of a plan to make a button for a tire manufacturer's website. In the space that follows, note coordinates and lengths you will use in the code.

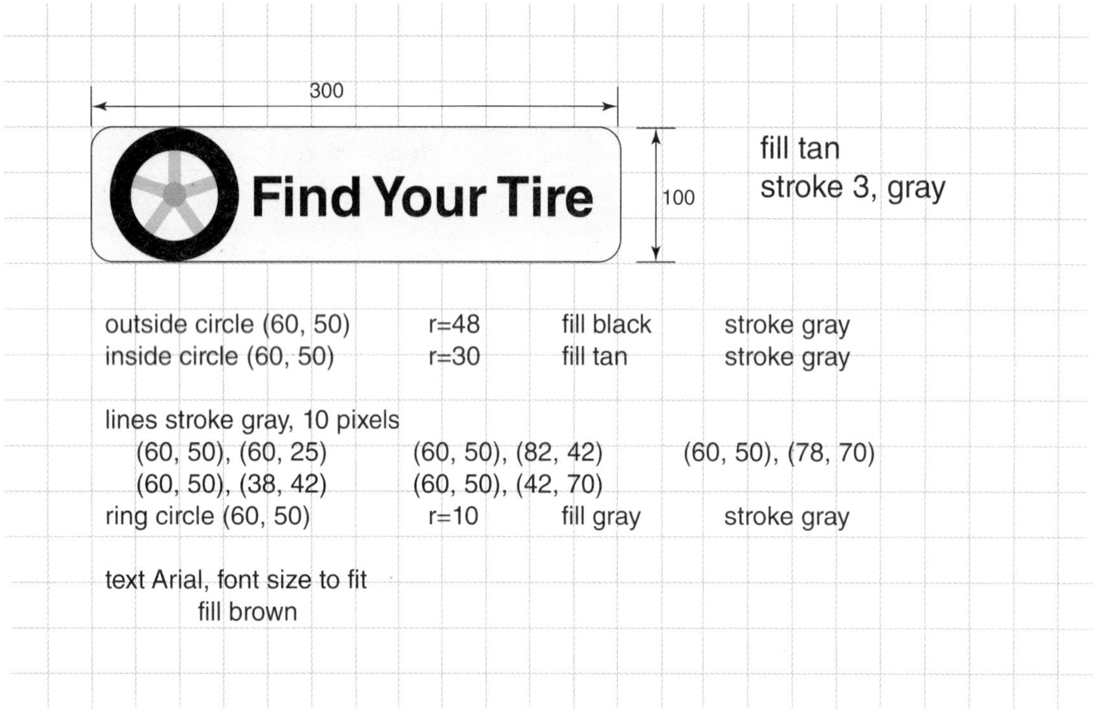
300
Find Your Tire
100
fill tan
stroke 3, gray
outside circle (60, 50) r=48 fill black stroke gray
inside circle (60, 50) r=30 fill tan stroke gray
lines stroke gray, 10 pixels
(60, 50), (60, 25) (60, 50), (82, 42) (60, 50), (78, 70)
(60, 50), (38, 42) (60, 50), (42, 70)
ring circle (60, 50) r=10 fill gray stroke gray
text Arial, font size to fit
fill brown

Goodheart-Willcox Publisher

2. Write an algorithm of how you plan to implement your design. In the sample, a rounded rectangle was defined, the tire was drawn as layered circles with lines for spokes. Text was added at the end.

___

___

___

___

___

___

3. Launch jGRASP, and open JavaFXLogoButtonSnippet.java. Locate comment 1, and draw the rounded rectangle for the button base according to your sketch. Use the coordinates and lengths on your sketch. Compile and run the application. Describe the results in the space provided.

___

___

___

___

4. Locate comment 2, and add code to draw the logo. Use coordinates relative to the rectangle.

   ```
   final double X = button.getX( );
   final double Y = button.getY( );
   ```

   Look at the sample drawing in step 1. The outside circle has a center of (60, 50) relative to the edges of the rectangle. Draw the circle so it is inside the rectangle by using this statement:

   ```
   Circle outside = new Circle( X + 60, Y + 50, 48 );
   ```

   The constants have been defined for you in the snippet. Choose one of your shapes and write the definition in the space below using the constants *X* and *Y*. When you have practiced this, complete the definition of the shapes making up your logo in the code. Remember to add all the shapes to the effects list.

   ___

   ___

   ___

   ___

5. Locate comment 3, and add code for the text of your button. Set the coordinates relative to the rectangle by using *X* and *Y* in the definition of the text object. Set the font and the fill properties.

6. Compile and run the application. If your result varied from the design, explain why you made changes in the space below. Submit your Java file and the design on grid paper.

   ___

   ___

   ___

   ___

Reflections

1. Explain what happens if the web page developer moves the button around on the app. What does defining constants for *X* and *Y* do for you and the developer?

2. Think about changing the size of the button. Explain the changes that would need to be made to the logo and text.

Debugging Challenge

Consider the following faulty code segment.

```
31  /***** 2. Define a Line object to draw a line in the middle of the domino.
32            Set the stroke to GRAY and the stroke width to 3.
33            Define the coordinates to scale with the coordinates of the
34            rectangle.
35  */
36  double lineY = domino.getY( ) + domino.getHeight( ) / 2;
37  double lineX1 = domino.getX( ) + 10;
38  double lineX2 = domino.getX( ) + domino.getWidth( ) - 10;
39  Line line = new Line( lineX1, lineY, lineX2, lineY );
40  line.setStroke( 3 );
41  line.setStrokeColor( Color.GRAY );
```

When compiled, the following errors are reported:

```
JavaFXDominoesAnswer.java:40: error: incompatible types: int cannot be converted to Paint
    line.setStroke( 3 );
                    ^
JavaFXDominoesAnswer.java:41: error: cannot find symbol
    line.setStrokeColor( Color.GRAY );
        ^
  symbol:   method setStrokeColor(Color)
  location: variable line of type Line

2 errors
```

How can the code be fixed? Record your response in the space that follows.

Name _____ Date _____ Class _____

CHAPTER 8
Selection

Selection statements allow the programmer to choose which statements to execute based on the value of data. Java provides equality, relational, and logical operators to use in building the tests. Selection can be made using an if-else statement or a switch statement.

Chapter Highlights

- A selection statement is a type of control-flow statement used to allow the program to choose a set of statements to execute based on a certain condition that evaluates to either true or false using equality, relational, and logical operators.
- The equality operators are == (equal to) and != (not equal to).
- The relational operators are < (less than), <= (less than or equal to), > (greater than), and >= (greater than or equal to).
- The logical operators are && (**AND**), || (**OR**), and ! (**NOT**).
- The basic selection statement in Java is an if-then statement, but there are variations, including if-then-else and if-then-else-if.
- The switch statement tests the value of a variable and selects statements based on any number of choices, and the break statement is used to interrupt the processing of cases in a switch statement.
- The scope of a variable is the block of code where it is defined, which may be as broad as at the method level or as limited as at the selection-statement level.
- A nested selection statement is a selection statement placed inside of another selection statement.
- De Morgan's laws and truth tables help simplify writing compound conditions.
- Comparing the data in objects is accomplished using the equals method.

While studying, look for the activity icon 📲 for:
- Vocabulary terms with e-flash cards and matching activities.
- Starter files for lab activities.

These activities can be accessed at www.g-wlearning.com/informationtechnology/1773

Warm-Up Exercises

1. Assume the following variables have been declared.
   ```
   boolean t = true;
   boolean f = false;
   boolean result;
   ```
   What is the value that will be assigned to the variable result after each of the following statements is executed?

   A. result = t && f; _____

   B. result = t || f; _____

   C. result = !t; _____

   D. result = t || f && 2 < 1 _____

   E. result = (t || f) && 2 < 1 _____

   F. result = t && !f; _____

   G. result = t && !(f || 2 < 1)_____

   H. result = f && !(f && 2 < 1)_____

_____ 2. Applying De Morgan's laws, select the expression that is equivalent to this expression:
   ```
   a < b && c != b
   ```
   A. a < b && c == b
   B. a > b && c == b
   C. a >= b || c == b
   D. a > b || c = b
   E. a < b && !(c != b)

_____ 3. Applying De Morgan's laws, select the expression that is equivalent to this expression:
   ```
   a > b || c == b
   ```
   A. a < b && c != b
   B. a > b && c == b
   C. a >= b || c == b
   D. a >= b && c != b
   E. a < b && !(c != b)

_____ 4. Consider the following code segment. What is printed as a result of executing this code segment?

```
int a = 10;
if ( a >= 10 ) {
   System.out.println( "Option 1" );
} else {
   System.out.println( "Option 2" );
}
```

A. Option 1

B. Option 2

C. Option 1
   Option 2

D. a is 10

E. Nothing is printed.

_____ 5. Consider the following code segment. What is printed as a result of executing this code segment?

```
int b = 20;
if ( b < 20 ) {
   System.out.println( "Option 1" );
} else {
   System.out.println( "Option 2" );
}
```

A. Option 1

B. Option 2

C. Option 1
   Option 2

D. b is 20

E. Nothing is printed.

_____ 6. Consider the following code segment. What is printed as a result of executing this code segment?

```
int a = 10;
int b = 20;
if ( a < 20 ) {
   System.out.println( "Option 1" );
   if ( b == 20 ) {
      System.out.println( "Option 2" );
   }
} else {
   System.out.println( "Option 3" );
}
```

A. Option 1

B. Option 2

C. Option 1
   Option 2

D. Option 2
   Option 3

E. Option 1
   Option 3

_____ 7. Consider the following code segment. What is printed as a result of executing this code segment?

```
int a = 10;
int b = 20;
if ( a < 20 ) {
   System.out.println( "Option 1" );
} else if ( b == 20 ) {
   System.out.println( "Option 2" );
} else {
   System.out.println( "Option 3" );
}
```

A. Option 1

B. Option 2

C. Option 1
   Option 2

D. Option 2
   Option 3

E. Option 1
   Option 3

_____ 8. Consider the following code segment. What is printed as a result of executing this code segment?

```
int a = 10;
int b = 20;
if ( a < 10 ) {
   System.out.println( "Option 1" );
} else if ( b > 20 ) {
   System.out.println( "Option 2" );
} else {
   System.out.println( "Option 3" );
}
```

A. Option 1

B. Option 2

C. Option 3

D. Option 2
   Option 3

E. Option 1
   Option 3

_____ 9. Consider the following code segment. What is printed as a result of executing this code segment?

```
String productType = "blouse";
switch ( productType ) {
   case "dress":
   case "blouse":
      System.out.println( "Type 1" );
      break;
   case "tie":
      System.out.println( "Type 2" );
   default:
      System.out.println( "Unknown type" );
}
```

A. Type 1
B. Type 2
C. Unknown type
D. Type 1
   Type 2
E. Type 2
   Unknown type

_____ 10. Consider the following code segment. What is printed as a result of executing this code segment?

```
String productType = "pants";
switch ( productType ) {
    case "dress":
    case "blouse":
      System.out.println( "Type 1" );
      break;
    case "tie":
      System.out.println( "Type 2" );
      break;
    default:
      System.out.println( "Unknown type" );
}
```

A. Type 1
B. Type 2
C. Unknown type
D. Type 1
   Type 2
E. Type 2
   Unknown type

Lab 8-1

Conditions

In this lab, you will use the relational, equality, and logical operators for building conditions. Before beginning this activity, download the files for this lab from the student companion website.

Learning Goals
- Construct conditional expressions using equality, relational, and logical operators.
- Diagram operator precedence.

Materials

- jGRASP Integrated Development Environment
- Starter files from the student companion website

Application and Extension of Knowledge

Using Comparisons to Determine Eligibility for a Driver's License

In this textbook chapter, there is a discussion about Glenn being able to get a driver's license if he is at least 16 years old. However, the rules vary by state. For example, for a person to get a driver's license in Florida, the person must be at least 15 years old. However, if the person is not yet 18 years old, he or she must have a parent's permission in the form of a Parental Consent Form. In this activity, you will write a program that inputs an age and whether or not the person has a Parental Consent Form. Your program should output true if the person is eligible for a driver's license or false if the person is not eligible.

Procedure

Often, the order in which the programmer receives the specifications for a program is not the clearest way to state the condition. For example, if you follow the order given above, the condition would be:

```
person is 15 or older AND ( person is >=18 OR has consent form )
```

Another way to state the condition is to first qualify persons 18 or older, then combine the two conditions necessary for persons who are 15–17 years old. The condition then becomes:

```
person is 18 or older OR ( person is 15-17 AND has consent form )
```

Begin with the **DriverLicenseSnippet.java** file. You will see the code for reading the person's age as an int and the prompt for reading if the person has a Parental Consent Form, also an int, is already written for you. The person should enter a 1 if he or she has a consent form and 0 if not.

1. Follow this algorithm, and plan your solution in the space provided.

   ```
   /***** 1. Define a condition for being 18 or older.
   */
   ```

   _____

   _____

   ```
   /***** 2. Define a condition for being aged 15-17.
   */
   ```

   _____

   _____

   ```
   /***** 3. Define a condition for having a consent form.
   */
   ```

   _____

   _____

```
/***** 4. Using logical operators, combine these conditions into an expression that
            evaluates to true or false to determine whether or not the person is eligible
            for a driver's license.
*/
```

_____

_____

_____

```
/***** 5. Output the result in a sentence.
*/
```

_____

_____

_____

2. Launch jGRASP, and open DriversLicenseSnippet.java starter file.
3. Enter the code you wrote above.

Reflections

1. What test values should you use to verify this code is correct? Be sure to test each of the conditions that would qualify a person for a driver's license. Also, test conditions that disqualify a person from getting a driver's license. Fill this table with your test values and expected results.

Tested Condition	Test Values	Expected Result

2. Research the regulations for qualifying for a driver's license in your state (or another state if you live in Florida). State those regulations and create a Boolean expression for determining eligibility for a driver's license in the state.

_____

_____

_____

Using Comparisons to Determine If a Year Is a Leap Year

Any leap year is evenly divisible by 4. However, not every year that is divisible by 4 is a leap year. For example, a century year is not a leap year, unless the year is also divisible by 400. This means that 1800 and 1900 are not leap years, but 2000 is a leap year. In this activity, you will create a program that accepts a year input and outputs true if the year is a leap year or false if it is not.

As you can see, multiple conditions must be met to determine that a year is a leap year. Rather than trying to deal with all conditions at once, break the large problem into smaller, more easily solved problems. This can be done by assigning each leap year determining factor to a boolean variable. Then the boolean variables can be combined using logical operators to get the final answer.

Begin with the LeapYearSnippet.java file. You will see the prompt for the user to enter the year as an int is already coded for you.

Procedure

1. Follow this algorithm, and plan your solution in the space provided.

   ```
   /***** 1. Define a condition for a year being divisible by 4, and assign it to a boolean
              variable.
   */
   ```

   _____

   _____

   ```
   /***** 2. Define a condition for a year being a century year, and assign it to a boolean
              variable.
   */
   ```

   _____

   _____

   ```
   /***** 3. Define a condition for a year being divisible by 400, and assign it to a
              boolean variable.
   */
   ```

   _____

   _____

   ```
   /***** 4. Using logical operators, combine the three boolean variables to determine if
              the year is a leap year, and assign the result to a boolean variable. Consider
              that if a year is divisible by 400, it is automatically a leap year. Other
              years are leap years if they are divisible by 4, but not divisible by 100.
   */
   ```

   _____

   _____

   _____

2. Launch jGRASP, and open LeapYearSnippet.java starter file.
3. Enter the code you wrote above.

Reflections

1. What test values should you use to verify this code is correct? Be sure to test each of the conditions that would qualify a year as a leap year. Also, test conditions that disqualify a year from being a leap year. Fill this table with your test values and expected results.

Tested Condition	Test Value	Expected Result

2. The algorithm used in this lab seems complicated. The actual length of an Earth year is 365.2422 days. That is just under 365 1/4 days. Explain how this determines the algorithm used in this activity. What would happen if there was a leap year every year, even years that are centuries?

Debugging Challenge

You want to write a program to test if a person's temperature is normal. The average normal temperature is 98.6 degrees Fahrenheit. However, any temperature between 97.0 degrees and 99.0 degrees is still considered normal. In your program, you write this code:

```
System.out.print( "Enter your temperature: " );
double temperature = input.nextDouble( );
boolean isNormal = temperature >= 97.0 && <= 99.0;
```

When you compile, you receive this error:

```
error: illegal start of expression
    boolean isNormal = temperature >= 97.0 && <= 99.0;
                                              ^
```

What is wrong, and how can you fix the code? Record your response in the space provided.

Lab 8-2
Selection Statements

In this lab, you will create various forms of if-then statements and write a program using a switch statement. Before beginning this activity, download the files for this lab from the student companion website.

Learning Goals
- Select the appropriate form of the if-then statement for an algorithm.
- Explain the function of the switch statement.
- Assess the scope and visibility of a given variable.
- Inspect code for common errors with selection statements.

Materials
- jGRASP Integrated Development Environment
- Starter files from the student companion website

Application and Extension of Knowledge

Using Selection Statements to Rate Tornados

Since 2007, the National Weather Service (www.weather.gov) uses the Enhanced Fujita Scale to rate tornados according to wind gusts and level of damage. You will use selection statements to rate tornados based on this scale.

Rating	Three-Second Wind Gusts (in MPH)
0	65–85
1	86–110
2	111–135
3	136–165
4	166–200
5	Over 200

First, you need to decide which form of the if-then statement is appropriate for this program. Because the classifications are six mutually exclusive ranges, an if-else-if statement would be the appropriate structure to use. Note also that each rating starts at one MPH higher than the previous rating. Therefore, these ratings can then be rearranged according to their minimum wind speeds:

Rating	Minimum Three-Second Wind Gusts (in MPH)
5	201
4	166
3	136
2	111
1	86
0	65

You also need to deal with the user entering a wind gust less than 65 MPH, which has no rating. That error case can be handled as a final **else** statement.

Begin with the TornadoRatingSnippet.java file. You will see the prompt for the user to enter the wind speed as an int is already coded for you.

Procedure

1. Follow this algorithm, and plan your solution in the space provided.

   ```
   /***** 1. Prompt the user to enter the wind speed. Also, create an empty String named
              rating to hold the classification.
    */
   ```

   ```
   /***** 2. Write the if-else-if statement to assign the correct value to rating.
    */
   ```

   ```
   /***** 3. Output the rating.
    */
   ```

2. Launch jGRASP, and open TornadoRatingSnippet.java starter file.
3. Enter the code you wrote above.

Reflections

1. To test whether or not this code is correct, you need to try all the boundary values. In fact, you should run this program with at least 12 different values to verify that each classification is correct. Fill this table with your test values and expected results. The first one is filled in for you. You can put two boundary values in each row.

Tested Condition	Test Value	Expected Rating
Boundaries for 5, 4	201 200	5 4

2. This set of classifications is arranged from high to low wind speeds, and the comparisons were made with the >= relational operator. Suppose the if-else-if statement is rearranged to process the wind speeds from low to high. For example, start with finding "not classifiable." Explain how the approach would be different and what relational operator you would use in this case.

_____

_____

_____

Using Selection Statements to Display Hours of Operation

You want to be able to post the daily hours of operation for your business on your website. You have decided to set up your business to have these hours:

Day	Hours of Operation
Sunday	closed
Monday	closed
Tuesday	9 a.m.–5 p.m.
Wednesday	10 a.m.–9 p.m.
Thursday	9 a.m.–5 p.m.
Friday	9 a.m.–9 p.m.
Saturday	9 a.m.–5 p.m.

One way to write this program is to use a switch statement where the inputted day of the week is the expression for the switch. Notice in the table of daily hours that several sets of days have the same hours of operation and that on two days the business is closed. Remember that with a switch statement, multiple case statements can be used to execute the same code. That feature of the switch statement will be exploited in this program.

Begin with the OpenHoursSnippet.java file. You will see the prompt for the user to enter the day of the week is already coded for you.

Procedure

1. Follow this algorithm, and plan your solution in the space provided.

    ```
    /***** 1. Start the switch statement here with its open brace;
              also add the end brace at this time.
    */
    ```

    _____

    _____

    ```
    /***** 2. On both Sunday and Monday, you are closed.
              Write two case statements for Sunday and Monday.
              Below the case statements, output a message that "we are closed"
              and code a break.
    */
    ```

    _____

    _____

    _____

    ```
    /***** 3. Tuesday, Thursday, and Saturday have the same hours.
              Write three case statements for Tuesday, Thursday, and Saturday.
              Below the case statements, output a message that "we are
              open from 9 a.m. to 5 p.m." and code a break.
    */
    ```

    _____

    _____

    _____

    ```
    /***** 4. Wednesday and Friday have the same hours.
              Write two case statements for Wednesday and Friday.
              Below the case statements, output a message that "we are
              open from 10 a.m. to 9 p.m." and code a break.
    */
    ```

    _____

    _____

    _____

    ```
    /***** 5. In case the user enters a day other than those listed,
              code a default statement and output a polite error message.
    */
    ```

    _____

    _____

2. Launch jGRASP, and open the OpenHoursSnippet.java starter file.
3. Enter the code you wrote above.

Reflections

1. How would you test this code?

2. If the user either misspells the day of the week or does not enter the day with the first letter capitalized and the other letters lowercase, what will the program do? How could the program be redesigned so the user does not need to enter the day spelled out and, thus, avoid this possible error?

Debugging Challenge

Consider this code that asks a user for their preferred book format:

```java
System.out.println( "In what format do you want the book?" );
String format = scan.next( );

switch ( format ) {
  case "Hardback":
    System.out.println( "The hardcover price is $25.99." );
    break;
  case "Paperback":
    System.out.println( "The paperback price is $18.99." );
  case "Electronic":
    System.out.println( "The electronic version is $8.99." );
    break;
}
```

When the user enters **Paperback**, the program outputs:

```
The paperback price is $18.99.
The electronic version is $8.99.
```

What is wrong, and how can you fix the code? Record your response in the space provided.

Name _____

Lab 8-3

Helpful Conditions

In this lab, you will write a program that uses nested conditionals. Typically, nested conditions are helpful when more information is needed beyond the first input. Before beginning this activity, download the files for this lab from the student companion website.

Learning Goals
- Apply nested conditionals.

Materials
- jGRASP Integrated Development Environment
- Starter files from the student companion website

Application and Extension of Knowledge

Using Nested Conditions for a Bike Rental App

Your job is to create a kiosk for bike rentals. The program in the kiosk calculates the rental fees based on several factors. Here is the fee schedule:
- If rental begins at 3 p.m. or later, the rental fee is $5.00 regardless of how long the bike is rented.
- For all other times, the rental fee is $2.00 for each hour or part of an hour the bike is rented.
- Persons with a frequent-biker membership number get 20 percent off the hourly rental fee. There is no discount for the flat $5.00 fee.

Input the checkout and return times as hours and minutes using a 24-hour clock; that is, 1:00 p.m. is 13; 2:00 p.m. is 14, and so on. You can assume the exit time is later than the entry time and the rental checkout and return is on the same day. The program should output the bike rental fee.

Hints: Convert the entry and exit times to minutes. For example, if the user enters 9 for the hour and 15 for the minutes, convert that to 555 minutes (9 × 60 + 15). Then, you can calculate the elapsed time by subtracting the total entry minutes from the total exit minutes.

Procedure

1. Follow this algorithm, and plan your solution in the space provided. When you start an if statement, write the start and end braces. It is also helpful to label the end braces with a comment. This will make it easier to see where an if statement starts and ends.

```
/***** 1. Define constants for
          the starting time for the flat rate: 15
          the flat rate: 5
          the hourly rate: 2
          the frequent biker discount rate: 20%
          how many minutes are in an hour: 60
*/
```

_____

_____

_____

_____

_____

```
/***** 2. Prompt the user for the rental start time in hours and minutes.
 */
```

_____

_____

_____

```
/***** 3. If the rental start hour is 15 or after, quote the flat rate fee. Note that no
          further information is needed at this time.
 */
```

_____

_____

_____

```
/***** 4. The remainder of the program is in the else clause.
          Prompt the user for the return time in hours and minutes.
 */
```

_____

_____

_____

```
/***** 5. Convert checkout and return times to minutes.
 */
```

_____

_____

_____

```
/***** 6. Calculate elapsed time in minutes.
          Convert elapsed time to hours, and add an hour for any minutes left over.
 */
```

_____

_____

_____

```
/***** 7. Prompt the user for the frequent-biker number or 0 if no number.
          If a frequent biker, calculate the rental fee with the discounted rate;
          otherwise, calculate the rental fee with the regular rate.
 */
```

_____

_____

_____

_____

2. Launch jGRASP, and open the BikeRentalSnippet.java starter file.
3. Enter the code you wrote above.

Name _____   Chapter 8 Selection **111**

Reflections

1. How would you test this code? Fill in the table below with the conditions to test and your test values.

Tested Condition	Test Values	Expected Fee

2. Explain the advantages to converting all times to minutes rather than calculating elapsed time using hours and minutes.

_____

_____

_____

Debugging Challenge

Consider this code that is intended to check the values of *x* and *y*:

```
int x = 0;
int y = 9;

if ( x == 0 ) {
   if ( y == 9 ) {
      System.out.println( "x is 0 and y is 9." );
   }
   else {
      System.out.println( "x is not 0 ." );
   }
}
```

When compiled, you get this compiler error on the last line of the .java file:

```
error: reached end of file while parsing
```

What is wrong, and how can you fix the code? Record your response in the space provided.

_____

_____

_____

_____

_____

Lab 8-4

Comparing Objects

In this lab, you will write programs that input Strings. You will need to use the equals( ) or equalsIgnoreCase( ) methods to check the values of these input Strings. Before beginning this activity, download the files for this lab from the student companion website.

Learning Goals
- Compare **Strings** using the **equals( )** and the **equalsIgnoreCase( )** methods.
- Compare numbers to determine output based on age.

Materials
- jGRASP Integrated Development Environment
- Starter files from the student companion website

Application and Extension of Knowledge

Coding the All-Knowing Wizard

In this activity, you will create an All-Knowing Wizard who can answer any question the user poses. This may seem beyond the scope of this course, and frankly, it is. Therefore, you will simulate this wizard by responding with a predefined answer depending on the first word of the question: who, where, when, how, or why. In other words, whenever the user asks a question beginning with the word, "Who," you might respond, "Mickey Mouse." You will take the same approach for each of the other question words.

Procedure

1. Follow this algorithm, and plan your solution in the space provided.

   ```
   /***** 1. Prompt for the question using Scanner's next method. Although the user can
              add a whole sentence before pressing the [Enter] key, only the first word is
              important.
   */
   ```

   ```
   /***** 2. Respond to the question by checking which word the user entered.
              Use equalsIgnoreCase so that the user can enter the word with uppercase or
              lowercase. Check for the words: "who", "where", "when", "how" and "why".
              Also check for input that does not match any of these words
              and respond that the wizard does not understand.
   */
   ```

2. Launch jGRASP, and open the WizardSnippet.java starter file.
3. Enter the code you wrote above.

Reflections
1. How would you test this code?

2. The program asks Scanner for only the first word. What happens to the rest of the sentence?

Comparing Values to Determine Output

Your job is to write a program that sells one ticket to a Segway tour. The ticket price is keyed to the visitor's age:
- Youth ticket (ages 12–20): $15
- Adult ticket (ages 21–64): $30
- Senior ticket (ages 65 and over): $20

The visitor must be at least 12 to take the tour. Also, no tours run on Monday. The output of this program should be one of the following:
- ticket price
- message that the visitor is too young to take the tour
- message that the tour does not operate on Monday
- message that the input day is not a valid day

Procedure

1. Follow this algorithm, and plan your solution in the space provided.

   ```
   /***** 1. Define constants for the prices.
   */
   ```

   ```
   /***** 2. Prompt the user for day of the tour. If the user enters Monday,
             output a message that the tour does not operate on Monday and code an else
             statement.
   */
   ```

   ```
   /***** 3. Inside the else, check that the day entered is a valid day.
             Use De Morgan's laws to form the correct condition.
             If not a valid day, output a message that the day is not valid
             and code an else statement.
   */
   ```

   ```
   /***** 4. At this point, the day is valid and is not Monday.
             Prompt the user for his or her age.
   */
   ```

```
/***** 5. If the user is under 12, output a message that he or she cannot tour.
         Otherwise, using an if-else-if, output the correct price according to the
         user's age.
*/
```

2. Launch jGRASP, and open the SegwayTourSnippet.java starter file.
3. Enter the code you wrote above.

Reflections

1. How would you test this program? Fill in the table below with the conditions to test and your test values.

Tested Condition	Test Values	Expected Fee

2. Explain how De Morgan's laws help in forming the condition to verify the day is valid.

Debugging Challenge

Consider this code that asks the user if he or she wants to roll a die:

```
Scanner input = new Scanner( System.in );
Random rand = new Random( );
System.out.println( "Do you want to roll the die? " );
String answer = input.next( );
if ( answer == "yes" ) {
   System.out.println( "The roll is " + rand.nextInt( 6 ) + 1 );
} else {
   System.out.println( "You chose not to roll the die." );
}
```

There are no compiler errors, but when you run the code and the user enters **yes**, you get this output:

```
"You chose not to roll the die."
```

What is wrong, and how can you fix the code? Record your response in the space provided.

Name _____ Date _____ Class _____

CHAPTER 9
Repetition

Repetition allows a program to repeat statements as many times as needed. A condition controls the number of times the code repeats. Repetition is useful for many purposes, but especially for reading and processing data from a file.

Chapter Highlights

- A loop is a block of code that is repeated as many times as required by the looping condition. Each time the block is repeated is called an iteration.
- The for loop iterates a block of code a predetermined number of times. This type of loop is ideal when the number of iterations is known before the loop begins.
- The while loop iterates a set of statements as long as a boolean expression is true. This type of loop is ideal when the number of iterations is not known before the loop begins.
- The do/while loop iterates a set of statements as long as a boolean expression is true. This type of loop is ideal when the statements in the loop need to be executed at least once.
- An infinite loop is one that iterates forever. This occurs when the loop's controlling condition never becomes false.
- To read the data in an external file, first import the java.io and java.util packages, then add throws FileNotFoundException to the main method, open the file, and read the file using a while loop controlled by the Scanner class hasNext method.

While studying, look for the activity icon 📲 for:
- Vocabulary terms with e-flash cards and matching activities.
- Starter files for lab activities.

These activities can be accessed at
www.g-wlearning.com/informationtechnology/1773

Warm-Up Exercises

_____ 1. Consider the following code segment. What is printed as a result of executing this code segment?

```
int s = 0;
for ( int i = 1; i < 4; i++ ) {
   s += i;
}
System.out.println( s );
```

A. 10
B. 0
C. 6
D. 15
E. 3

_____ 2. Consider the following code segment. What is printed as a result of executing this code segment?

```
int c = 0;
for ( int i = 1; i <= 5; i++ ) {
   c++;
}
System.out.println( c );
```

A. 10
B. 5
C. 4
D. 0
E. 1

_____ 3. Consider the following code segment. What is printed as a result of executing this code segment?

```
int b = 0;
for ( int i = 0; i < 5; i += 2 ) {
   b++;
}
System.out.println( b );
```

A. 3
B. 5
C. 4
D. 10
E. 15

_____ 4. Consider the following code segment. What is printed as a result of executing this code segment?

```
double a = 0;
int c = 0;
for ( int i = 0; i <= 5; i++ ) {
   a += i;
   c++;
}
System.out.println( a / c );
```

A. 2.0
B. 15
C. 5.0
D. 2.5
E. 5.5

For questions 5 and 6, consider the following code segment.

```
String o = "";
System.out.print( "Enter a number: " );
int n = input.nextInt( );
while ( n != 0 ) {
   o +=  n + " ";
   System.out.print( "Enter a number: " );
   n = input.nextInt( );
}
System.out.println( o );
```

_____ 5. If the user enters 1 at the first prompt, 2 at the second, 3 at the third, and 0 at the fourth, what is printed as the last line as a result of executing this code segment?

A. 1 2 3
B. 1 2 3 0
C. 0
D. 0 3 2 1
E. Nothing is printed.

_____ 6. If the user enters 0, what is printed as the last line as a result of executing this code segment?

A. 1 2 3
B. 1 2 3 0
C. 0
D. 0 3 2 1
E. An empty line.

7. Consider the following code segment. What is printed as a result of executing this code segment?

    ```
    int x = 0;
    while ( x < 4 ) {
       System.out.print( x + " " );
       x++;
    }
    ```

    A. 1 2 3 4
    B. 1 2 3
    C. 0 1 2 3
    D. 0 1 2 3 4
    E. 0 2 4

8. Consider the following code segment. What is printed as a result of executing this code segment?

    ```
    int x = 4;
    while ( x > 0 ) {
       System.out.print( x + " " );
       x--;
    }
    ```

    A. 4 3 2 1 0
    B. 4 3 2 1
    C. 3 2 1 0
    D. 1 2 3 4
    E. 0 1 2 3 4

9. Consider the following code segment. What is printed as a result of executing this code segment?

    ```
    int a = 6;
    do {

       a *= 2;

    } while ( a < 24 );
    System.out.println( a );
    ```

    A. 6
    B. 12
    C. 3
    D. 18
    E. 24

_____ 10. Consider the following code segment. What is printed as a result of executing this code segment?

```
int a = 6;
do {

   a *= 2;

} while ( a < 6 );
System.out.println( a );
```

A. 6
B. 12
C. 3
D. 18
E. 24

Lab 9-1

Java Loops

In this lab, you will write for loops, while loops, and do/while loops. Before beginning this activity, download the files for this lab from the student companion website.

Learning Goals
- Apply counter-controlled for loops in Java.
- Create condition-controlled while loops in Java.
- Design condition-controlled do/while loops in Java.

Materials
- jGRASP Integrated Development Environment
- Starter files from the student companion website

Application and Extension of Knowledge

Using a For Loop to Calculate Fibonacci Numbers

The Fibonacci sequence is a series of numbers formed by adding the sum of the two previous numbers. The first number in the series is defined to be 0, and the second number is defined to be 1. Beginning with these two numbers, the series continues as: 0, 1, 1, 2, 5, 8, 13, 21, and so on. The table below shows how the sequence is determined.

Term	0	1	2	3	4	5	6	7	8	...
Value	0	1	1	2	3	5	8	13	21	...
Sum	—	—	0 + 1	1 + 1	2 + 1	2 + 3	3 + 5	5 + 8	8 + 13	...

In this activity, you will create an application that asks the user for a target term in the series and then calculates the value of the Fibonacci number for that term.

A viable approach to this problem is to start at 2 and from there calculate the series of Fibonacci numbers until reaching the user's target. Because you will know the number of iterations, a for loop is a logical choice for the loop.

A challenge to this problem is how to determine the value of the previous two numbers at any point so they can be added to create the next number in the series. This can be done by saving the previous numbers as each new term is calculated.

Procedure

1. Follow this algorithm, and plan your solution in the space provided.

    ```
    /****** 1. Input the user's desired term. Ask for a positive number.
    */
    ```

    ```
    /***** 2. If the user has entered 0 or 1, the answer is known.
              The value for the terms 0 and 1 is the same as the term.
              Output the term.
              Code an else statement for the remainder of the program.
    */
    ```

    ```
    /***** 3. For other values, calculation needs to be done.
              Initialize the two previous Fibonacci numbers
              to 0 (term 0 ) and 1 (term 1).
              Also define the result, fib, initialized to 0.
              Use the data type long because the numbers get large quickly.
    */
    ```

    ```
    /***** 4. Write the for loop. Start the counter at 2.
              Continue until reaching the target term, incrementing by 1.
              Inside the for loop, add the last two numbers to the result.
              Then, save the last number as the second previous number,
              and save the result as the last number.
    */
    ```

    ```
    /***** 5. Output the result in a sentence, formatted with commas every three digits.
    */
    ```

2. Launch jGRASP, and open FibonacciSnippet.java starter file.
3. Enter the code you wrote above.
4. Test the code.

Reflections

1. What happens if the user enters a negative number? How can you ensure the user enters a positive number?

2. Research the history of the Fibonacci numbers, and give some examples of its applications.

Using a While Loop to Calculate the Total Price of a Market Basket

In this activity, you will write a program to simulate a self-checkout machine at a big-box store. Your program will calculate the total cost of a customer's items. The customer will enter the price for each item. Each customer could have a different number of items, so when the program begins, how many items this customer will have is unknown. The customer will need to give the program a signal that there are no more items to scan.

This type of application is well-suited to a while loop. A while loop continues until some signal indicates that there is no more data to process.

Procedure

1. Follow this algorithm, and plan your solution in the space provided.

    ```
    /***** 1. Initialize the total price, a double, to 0.0.
    */
    ```

    ```
    /***** 2. Perform the priming read to "scan" the first price
    */
    ```

    ```
    /***** 3. Write the while loop. Inside the loop, add the item price
             to the total and "scan" the next item price.
             The loop ends when the user enters 0.0 for the price.
    */
    ```

```
/***** 4. Output the final price.
          Input the amount of money the user pays.
          If the user does not pay enough, keep prompting the user until the user pays at
          least the total amount, and then print a message with the change.
          Note that since we do not know how many times the user will insert money
          before the total is paid, this is another while loop.
 */
```

2. Launch jGRASP, and open SelfCheckoutSnippet.java starter file.
3. Enter the code you wrote above.
4. Test the code.

Reflections

1. Explain the purpose of the priming read before the while loop begins.

   _____
   _____
   _____

2. Explain the purpose of the update read inside the while loop.

   _____
   _____
   _____

Using a Do/While Loop to Calculate a Square Root

The Math class in the Java Class Library provides a method for calculating the square root of a number. Although this method is convenient to use, it is also good to know how to write Java code to calculate a square root without using the Math.sqrt method. In this activity, you will write a program to calculate the square root using the Babylonian method. This method uses a loop to approximate a square root of a positive number by guessing values until the new guess is sufficiently close to the previous guess. Here are the steps to the algorithm:

1. Make two variables to hold the old and new guesses. Initialize the new guess to be the positive number you entered divided by 2.
2. Save the new guess as the old guess. Then create a new guess by applying the formula:
   new guess = (old guess – number ÷ old guess) ÷ 2
3. Repeat step 2 until the absolute value of the new guess and old guess are sufficiently close. Usually, a small threshold is chosen as a tolerance level for the difference.

You will need to iterate the loop until a condition is met. That is, until the old and new guesses are sufficiently close. This means a condition-controlled loop is appropriate. This leaves a choice between a while loop and a do/while loop. You know there needs to be at least one guess, so the loop must be executed at least once. In this situation, where a loop will be executed at least once, a do/while loop is the best option.

Procedure

1. Follow this algorithm, and plan your solution in the space provided.

   ```
   /***** 1. Define a constant THRESHOLD as .0001 for the difference tolerance.
             Initialize oldGuess to 0.
             Initialize newGuess to number / 2.
   */
   ```

   ```
   /***** 2. In a do/while loop,
             1. assign the newGuess to oldGuess
             2. apply the formula for calculating a new guess
             End the loop when the absolute value between the new and old guesses
             is less than the threshold defined above.
   */
   ```

   ```
   /***** 3. Output the square root.
   */
   ```

2. Launch jGRASP, and open SquareRootSnippet.java starter file.
3. Enter the code you wrote above.
4. Test the code.

Reflections

1. How can you test this program?

2. Is there any circumstance that would cause the first guess to be correct?

Debugging Challenge

In your program, you write this code to output the message Greetings! three times:

```
for ( int i = 1; i <= 3; i-- ) {
   System.out.println( "Greetings!" );
}
```

The code compiles with no errors, but when you run the code, the message is printed forever. What is wrong, and how can you fix the code? Record your response in the space provided.

_____

_____

_____

_____

_____

Lab 9-2

Applying Loops

In this lab, you will read and analyze data from an external file. Specifically, you will find an average value, minimum and maximum values, and counts. Before beginning this activity, download the files for this lab from the student companion website.

Learning Goals
- Read data from a text file.
- Calculate an average.
- Find minimum and maximum values.
- Count data that meet some criteria.

Materials
- jGRASP Integrated Development Environment
- Starter files from the student companion website

Application and Extension of Knowledge

Reading Data from a Bank Account Statement

In this activity, you will write a program that reads data from a file and performs some calculations and analysis on the data. The file contains transactions performed on a bank account over the last month. Your program should read the file and apply the transactions to the balance. Deposits should be added to the balance, and withdrawals should be subtracted from the balance. After reading the file, your program should print the average balance, the highest balance, the lowest balance, and the number of deposits and withdrawals for the month. For each transaction, your program should output the type of transaction, the amount of the transaction, and the new balance.

The name of the file your program will read and analyze is BankTransactions.txt. The first line of the file contains three pieces of data, separated by white space:
- month for which the transactions apply (a String)
- account number (a String)
- beginning balance (a double)

All lines after the first line have two pieces of data, separated by white space:
- transaction type (a String), the value will be either "Deposit" or "Withdrawal"
- amount deposited or withdrawn (a double)

It is recommended that you write this program in stages. Get each stage working, then add code for other functionality.

1. Read the file and apply the transactions.
2. Add code to count the deposits and withdrawals.
3. Add code to find the average balance.
4. Add code to find the minimum and maximum balances.

Writing code in stages like this is called *stepwise refinement* and makes writing and debugging code easier.

Start with the BankStatementSnippet.java file. Notice that the phrase throws FileNotFoundException is already added to the end of the main method. Notice also that the java.io and java.util packages are already imported.

Procedure

1. Follow this algorithm, and plan your solution in the space provided.

   ```
   /***** 1. Open the BankTransactions.txt file by creating a File object
             and a Scanner object to read the file.
   */
   ```

   ```
   /***** 2. Read the first line of the file and output a header for the statement:
             Bank Statement for <month>
             Account number: <account number>    Beginning balance: <balance>
             Then output a header for the transactions:
             Transaction     Amount($)      Balance($)
   */
   ```

```
/***** 3. Write a while loop to read the file.
         In the while loop, read the transaction type and the amount
         from the next line in the file.
         If the transaction is "Deposit", add the amount to the balance.
         If the transaction is "Withdrawal", subtract the amount from the balance.
         Remember that these are Strings, so the equals method should be used.
         Output a line with the transaction type, the amount, and the new balance.
```

_____

_____

_____

_____

_____

_____

_____

_____

_____

_____

```
/***** 4. End the while loop. As you add more functions to the program,
         add appropriate output statements after the loop ends to report
         your answers.
*/
```

_____

_____

_____

2. Launch jGRASP, and open BankStatementSnippet.java starter file.
3. Enter the code you wrote above.
4. Test that code.
5. Add functionality to count the number of deposits and withdrawals. You will need to initialize two counters to 0 before the while loop begins. In the loop, where your code determines the transaction type, add 1 to the appropriate counter.
6. Test that code.
7. Add functionality to calculate the average balance. Initialize a total balance before the while loop begins. In the loop, add each new balance to the total. When the loop is finished, divide the total by the sum of the deposits and withdrawals counts.
8. Test that code.
9. Add functionality to find the minimum and maximum balances. Initialize the minimum and the maximum to the beginning balance before the while loop begins. In the loop, compare the new balance to the minimum. If the new balance is lower, replace the minimum with the new balance. Also, in the loop, compare the new balance to the maximum. If the new balance is higher, replace the maximum with the new balance.
10. Test that code.

Reflections

1. Why is it a good practice to use stepwise refinement in writing this program? Discuss your experiences using this method to develop this program.

   _____
   _____
   _____

2. What other transaction types are possible in a bank statement? Name transaction types other than deposits and withdrawals, and explain how you would change the program to handle those transaction types.

   _____
   _____
   _____

Debugging Challenge

Consider this code that is intended to read the file Input.txt, which contains integers.

```java
import java.io.*;
import java.util.*;

public class ReadIntegers {
   public static void main( String [ ] args ) {
      File inputFile = new File( "Input.txt" );
      Scanner file = new Scanner( inputFile );

      while ( file.hasNext( ) ) {
         int number = file.nextInt( );
      } // end while

   } // end main
} // end class
```

When you compile, you get this compiler error on the last line of the Java file:

```
error: unreported exception FileNotFoundException; must be caught or declared to be thrown
     Scanner file = new Scanner( inputFile );
                    ^
```

What is wrong, and how can you fix the code? Record your response in the space provided.

_____
_____
_____
_____
_____

Notes

Name _____ Date _____ Class _____

CHAPTER 10

String Processing

Strings are an important part of programs. A Java String is a sequence of characters. Strings are objects, rather than a primitive type such as ints or doubles. The String class is in the java.lang package so there is no import needed to use Strings.

Chapter Highlights

- A Java String is a sequence of characters and can be created with constructors, assigning a literal to a String reference, or as a return value from a method.
- The Scanner class in the java.util package contains two methods for inputting a String: next method and nextLine method.
- Strings can be concatenated to combine them.
- The length of a String is found using the length method of the String class, which counts the number of characters in the String.
- Characters in a String are identified by an index, which always begins at 0 and ends at one less than the length of the String.
- A substring is part of a String, and the substring method is used to extract characters from a String; an invalid index argument from this method will generate a StringIndexOutOfBoundsException error.
- The case of all letter characters in a String can be changed using the toUpperCase and toLowerCase methods.
- The indexOf method will find the first character in a String that matches a specified case-sensitive search String; if the search String is not found, the method returns –1.
- The replace method is used to find characters in a String and replace them with different characters; the search String is case-sensitive.

While studying, look for the activity icon for:
- Vocabulary terms with e-flash cards and matching activities.
- Starter files for lab activities.

These activities can be accessed at
www.g-wlearning.com/informationtechnology/1773

- The charAt method can be used when traversing a String either forward or backward; the charAt method returns the character at a specified index, and this method can be used in conjunction with the next method to get input as a char data type.
- Primitive data types do not have methods, but a wrapper class, such as Character, encloses a primitive data type into an object that can then be used to call methods.

Warm-Up Exercises

_____ 1. Consider the following code segment. What is printed as a result of executing this code segment?

```
String holiday = "Fourth";
holiday = holiday + " of ";
holiday += "July";
System.out.println( holiday );
```

A. July Fourth
B. Fourth of July
C. FourthofJuly
D. 4th of July
E. July of Fourth

_____ 2. Consider the following code segment. If the user enters "Algebra and Trigonometry," what is printed as a result of executing this code segment?

```
System.out.print( "Enter your course name: " );
String courseName = input.next( );
System.out.println( courseName );
```

A. Algebra and Trigonometry
B. Trigonometry
C. Algebra and
D. Algebra
E. Nothing is printed.

_____ 3. Consider the following code segment. What is printed as a result of executing this code segment?

```
String test = "Testing 1 2 3.";
System.out.println( test.length( ) );
```

A. 7
B. 14
C. 11
D. Testing 1 2 3
E. Testing

_____ 4. Consider the following code segment. What is printed as a result of executing this code segment?

```
String test = "Testing 1 2 3.";
System.out.println( test.toUpperCase( ) );
```

A. TESTING 1 2 3
B. TEST
C. Testing 1 2 3
D. testing 1 2 3
E. TESTing 1 2 3

_____ 5. Consider the following code segment. What is printed as a result of executing this code segment?

```
String test = "Testing 1 2 3.";
System.out.println( test.substring( 3, 6 ) );
```

A. 1 2 3
B. stin
C. ting
D. tin
E. Test

_____ 6. Consider the following code segment. What is printed as a result of executing this code segment?

```
String test = "Testing 1 2 3.";
System.out.println( test.substring( 10 ) );
```

A. 1 2 3.
B. 2 3.
C. 3.
D. 2
E. Testing

_____ 7. Consider the following code segment. What is printed as a result of executing this code segment?

```
String test = "Testing 1 2 3.";
System.out.println( test.indexOf( " " ) );
```

A. 7
B. 7 9 11
C. 6
D. 8
E. 11

_____ 8. Consider the following code segment. What is printed as a result of executing this code segment?

```
String city = "Philadelphia";
System.out.println( city.replace( "p", "?" ) );
```

A. Philadelphia
B. ?hiladelphia
C. ?hiladel?hia
D. Philadel?hia
E. PHILADEL?HIA

_____ 9. Consider the following code segment. What is printed as a result of executing this code segment?

```
String alpha = "abcde";
System.out.println( alpha.charAt( 0 ) );
```

A. a
B. b
C. c
D. d
E. e

_____ 10. Consider the following code segment. What is printed as a result of executing this code segment?

```
String alpha = "abcde";
System.out.println( alpha.charAt( alpha.length( ) - 1 ) );
```

A. a
B. b
C. c
D. d
E. e

Lab 10-1

Creating Strings

In this lab, you will print a Java source file with the numbers added. Before beginning this activity, download the files for this lab from the student companion website.

Learning Goals
- Input Strings into a program.
- Create concatenated Strings.
- Pass a String as a parameter to a method.

Materials
- jGRASP Integrated Development Environment
- Starter files from the student companion website

Application and Extension of Knowledge

Creating Strings to Add Line Numbers to a Printout

In this activity, you will add line numbers to a printout. Specifically, you will read a file and output each line preceded by a line number. To do this, a counter variable is needed that keeps track of the current line number. This variable should be initialized to 1, then incremented by 1 after each line has been printed.

Begin with the AddLineNumbersSnippet.java file. You will see the appropriate Java packages for reading files have been imported, and that throws FileNotFoundException has been added to the main.

Procedure

1. Follow this algorithm, and plan your solution in the space provided.

```
/***** 1. Input the file name. A file name is one word,
          so the Scanner next method would be appropriate.
          Then, append a .java extension to the file name.
*/
```

_____

_____

```
/***** 2. Open the file name to prepare for reading the file.
*/
```

_____

```
/***** 3. Prepare for numbering the lines
          by defining a lineNumber variable. Set its
          value to the first line number: 1.
*/
```

_____

```
/***** 4. Set up the while loop to read the file as long as the Scanner
          hasNext method returns true. Code the while loop header
          as well as the opening and closing braces for the loop.
*/
```

_____

_____

_____

_____

```
/***** 5. Inside the loop, read a line. Because the line will potentially
         have multiple words, use the Scanner nextLine method.
         Then, output a line consisting of the line number with a width of 3,
         a space, and then the line. Do this by using String.format to
         create a three-character String consisting of the line number. Then,
         output the line by concatenating the line number, a space, and the line.
         Finally, add 1 to the linenumber variable.
*/
```

2. Launch jGRASP, and open AddLineNumbersSnippet.java starter file.
3. Enter the code for the algorithm you wrote above.
4. Test the code. When prompted for the file name, enter in the name of this file (AddLineNumbersSnippet) *without* the .java extension. Remember, the program adds .java to whatever file name is entered by the user.

Reflections

1. How would the body of the loop change if you initialized the lineNumber variable to 0?

2. This program assumes the file is a Java source file. It automatically adds the .java extension to whatever name the user enters. Explain an advantage and disadvantage of this assumption.

Debugging Challenge

In your program, you write this code to input the user's name:

```
Scanner input = new Scanner( System.in );

System.out.print( "Enter your first name: " );
String name = input.nextInt( );
```

When you compile the code, you get this error:

```
error: incompatible types: int cannot be converted to String
    String name = input.nextInt( );
                               ^
```

What is wrong, and how can you fix the code? Record your response in the space provided.

_____

_____

_____

_____

_____

Lab 10-2

String Methods

In this lab, you will convert an English phrase to the jargon known as pig latin. Before beginning this activity, download the files for this lab from the student companion website.

Learning Goals
- Find the length of a String.
- Determine the index of a character in a String.
- Extract a substring from a String.
- Search a String for characters.

Materials
- jGRASP Integrated Development Environment
- Starter files from the student companion website

Application and Extension of Knowledge

Rearranging Strings

In this activity, you will ask the user to input a phrase or sentence and translate that phrase to pig latin. Pig latin is a language game that is sometimes used between people to disguise what they are saying around people who do not know pig latin. Here is an example of an English phrase and its translation to pig latin:
- English phrase: This is my best work
- Pig latin: isThay isyay ymay estbay orkway

There are several variations in the rules for creating pig latin. In this activity, the following rules will be applied. For each word in the phrase or sentence:
- Leading consonants should be removed from the beginning of the word and appended to the end of the word followed by "ay".
- If a word begins with a vowel, simply append "yay" to the end of the word. Consider *Y* a vowel.
- Assume the phrase contains only letters, no special characters or punctuation.
- Assume only one space separates each word in the phrase.

This Lab has several challenges. First, how can you recognize where a word ends? You could search for a space and extract a substring from the beginning of the String to the space. This strategy should work for all words except the last word. Most likely, the user will press the [Enter] key immediately after the last word, so searching for a space for the last word will not work. You can get around this problem by first trimming the phrase to remove leading and trailing spaces. Then you can append a space to the end, thus guaranteeing that you will find the last word.

Another challenge in extracting words from the phrase is that the indexOf method will find only the first occurrence of the search string. You can manage that challenge by removing each word from the beginning of the phrase after you translate that word. The phrase will then contain only words that have not been translated.

To implement the algorithm, there needs to be a way to determine whether letters are vowels or consonants. One way to do this is to create a String consisting of all the vowels. You can then search to see if a character is in that String. If so, the character is a vowel. Otherwise, the character must be a consonant. Follow this algorithm for the program:

1. Prompt the user for the English phrase.
2. Trim leading and trailing spaces and append a space to the end.
3. Define a String containing the uppercase and lowercase versions of all the vowels, including the letter *Y*.
4. Define an empty String to hold the pig latin translation.
5. Find the index of a space in the English phrase.
6. While a space was found:
    A. Extract all characters before the space into a word.
    B. Create an empty String to hold the consonants found.
    C. While the next character is not found in the vowels String:
        1. Append the character to the consonants String.
        2. Remove the character from the word.
    D. If no consonants were found:
        1. Create the pig latin word by appending "yay"; else
        2. Create the pig latin word by appending the consonants and "ay".
    E. Add the pig latin word to the String for the pig latin phrase.
    F. Remove the word from the English phrase.
    G. Find the next word by searching for a space.
7. Output the pig latin phrase.

Before beginning the program, walk through this algorithm for the first two words of this example.
- English phrase: This is my best work
- Pig latin: isThay isyay ymay estbay orkway

For the first word, "This":
- pigLatin => ""
- vowels => "aAeEiIoOuUyY"
- consonants => ""
- word => "This"

Search for "T" and find it is not in vowels, so move "T" to a String containing consonants and remove "T" from the word:
- consonants => "T"
- word => "his"

Search for "h" and find it is not in vowels, so append "h" to the consonants String and remove "h" from the word:
- consonants => "Th"
- word => "is"

Search for "i" and find that it is in vowels. Now you are ready to create the pig latin word. Start with the remainder of word, append the consonants, and then append "ay".
- pigLatin => "isThay"

Remove "This" from the English phrase, extract the next word "is", and repeat the process. This time, there are not any leading consonants, so create the pig latin by appending "yay" to the word:
- consonants => ""
- word => "is"
- pigLatin = "isThay isyay"

Continue with this algorithm until the end of the sentence.

Procedure

1. Plan your solution in the space provided.

    ```
    /***** 1. Prompt the user for a phrase or sentence.
             Trim leading and trailing spaces,
             then add a space to the end.
    */
    ```

    ```
    /***** 2. Define a String containing all vowels, including Y.
             Include both uppercase and lowercase versions of the vowels.
             Also, define the empty String pigLatinPhrase.
             Each translated word will be appended to this String.
    */
    ```

```
/***** 3. Use a while loop to move through the phrase word by word.
          The condition will be that there is still another word to translate.
          You will know if more words are in the phrase by checking whether
          or not there is another space in the phrase. Prime the loop condition
          by searching for the first space. Then write the loop condition and the
          opening and closing braces. Also code the update search for the next
          space before the closing brace.
*/
```

___
___
___
___

```
/***** 4. Inside the loop, extract the letters before the space into the next word.
          Create an empty String to hold the leading consonants.
          Begin another while loop inside the outer while loop
          that will check whether or not the first letter is a consonant.
          This while loop's condition should be that the letter is
          not in the vowels String.
*/
```

___
___
___
___

```
/***** 5. In the inner while loop body, the current letter must be
          a consonant. Append the letter to the consonants String
          and remove the letter from the word.
*/
```

___
___
___
___

```
/***** 6. After the while loop ends, the leading consonants have been
          moved to the consonants String. The word must now start with
          a vowel and you are ready to create the pig latin word. If the
          consonants String is empty, the original word began with a vowel,
          so append "yay" to the word. Otherwise, append the consonants
          to word, plus "ay". Append the new word to the pigLatinPhrase String.
*/
```

___
___
___
___

```
/***** 7. Remove the translated word from the phrase.
          Be sure that the update for the outer while loop
          condition (searching for the next space) is the
          last statement in the loop.
*/
```

```
/***** 8. Output the pig latin translation.
*/
```

2. Launch jGRASP, and open PigLatinSnippet.java starter file.
3. Enter the code you wrote above.
4. Test that code.

Reflections

1. Which methods of the String class did you use in this program? What function did each method perform in the algorithm?

2. How should you test this code to thoroughly check it?

Debugging Challenge

Consider this code that is intended to print the word Java:

```
String course = "Intro to Java";
int indexOfJ = course.indexOf( "j" );
System.out.println( course.substring( indexOfJ ) );
```

The code compiles without error, but when you run the code, this exception is reported:

```
Exception in thread "main" java.lang.StringIndexOutOfBoundsException: String index out of range: -1
```

What is wrong, and how can you fix the code? Record your response in the space provided.

_____
_____
_____
_____
_____

Lab 10-3

Processing Strings Character by Character

In this lab, you will convert a binary number to a decimal number. Before beginning this activity, download the files for this lab from the student companion website.

Learning Goals

- Extract characters by traversing a String.

Materials

- jGRASP Integrated Development Environment
- Starter files from the student companion website

Application and Extension of Knowledge

Manipulating Strings to Convert Binary to Decimal

In this activity, you will write a program that converts a binary number input to the decimal equivalent of that number and outputs it. Prompt the user for a String representing a binary number; that is, it contains only 0s, 1s, and spaces. If the String contains other characters, reprompt the user until the String is valid. This can be done with a do/while loop and a flag. Then, convert that binary number (String) to its decimal equivalent. There are two loops here:

- Validate that all characters are 0, 1, or a space.
- Convert the String to a decimal number.

Remember the algorithm for converting a binary number to decimal: For each position in the binary number that is 1, add the corresponding power of 2 to the decimal number. The rightmost digit is 2^0 power. Moving to the leftmost digit, each digit represents a higher power of 2. For example, if the input is:

0001 1010

the corresponding powers of 2 are:

Powers of 2	2^7	2^6	2^5	2^4	none	2^3	2^2	2^1	2^0
Binary number	0	0	0	1		1	0	1	0

So, the decimal value is:
$= 2^1 + 2^3 + 2^4$
$= 2 + 8 + 16$
$= 26$

Notice that the 0 characters and spaces add nothing to the decimal value. Follow this algorithm:

1. Start with a decimal number of 0.
2. Define a variable that will keep the current power of 2, and initialize it to 0.
3. Use a for loop to process each character in the binary number (String), starting from the rightmost character. For each character:
   - If the character is a 1, then add the current power of 2 to the decimal number.
   - If the character is a 0 or 1, increment the power of 2.
   - If the character is a space, do nothing.

   Before beginning the program, walk through this algorithm with these characters: 10 01

   ```
   decimal => 0
   power of 2 => 0
   ```

for loop:

```
extract last character: 10 01 => 1
character is a 1, so raise 1 to the 0th power => 1
add 1 to decimal value:
   decimal => 0 + 1 => 1
increment the power of 2 => 1

extract next to last character 10 01  => 0
character is a 0, ignore
increment power of 2 => 2

extract character third from the end:  <space>
do nothing

extract character 4th from the end   10 10:  => 0
character is a 0, ignore
increment power of 2 => 3

extract character 5th from the end  (the first character) 10 01 = > 1
character is a 1, so raise 1 to the 3rd power => 8
add 8 to decimal value:
   decimal => 1 + 8 => 9
end of loop
decimal => 9
```

Procedure

1. Plan your solution in the space provided.

```
/***** 1. Define a String named binary and a boolean flag named validBinary.
 */
```

```
/***** 2. In a do/while loop, input the binary number as a String.
          Repeat until all characters are either 0, 1, or a space.
          Set the flag to true after reading the binary String, then set the flag to false
          if an invalid character is found. The do/while condition should check the flag.
 */
```

```
/***** 3. Begin to convert the number to decimal.
          Initialize the decimal number as 0.
          Initialize the power of 2 as 0.
 */
```

```
/***** 4. In a for loop, starting with the last character in the binary number String
          and moving to the first character, extract the character. If the character
          is a 1, raise 2 to the current power of 2, and add that value to the decimal
          number. If the character is 1 or 0, increment the power of 2. If the character
          is a space, do nothing.
*/
```

```
/***** 5. Output the decimal value.
*/
```

2. Launch jGRASP, and open BinaryToDecimalSnippet.java starter file.
3. Enter the code you wrote above.
4. Test that code.

Reflections

1. Discuss the advantages of starting at the rightmost digit of the binary String.

2. How can this conversion be done without the powerOf2 variable? In other words, how can you calculate the appropriate power of 2 for each position? Is there a relationship between i, the length of the String, and the number of spaces? Hint: start the for loop at the leftmost digit rather than the rightmost digit.

Debugging Challenge

Consider this code that is intended to print every character in a **String** separated by a space:

```
String course = "Intro to Java";
for ( int i = 0; i < course.length( ) - 1; i++ ) {
   System.out.print( course.charAt( i ) + " " );
}
System.out.println( );
```

The code compiles without error, but when you run the code, the course name is printed without its last letter:

I n t r o   t o   J a v

What is wrong, and how can you fix the code? Record your response in the space provided.

Name _____ Date _____ Class _____

CHAPTER 11

Managing Input and Output

Using Scanner and System.out.println( ), you can capture input and print results easily. However, the old adage "garbage in, garbage out" rings true if the input is not validated. This chapter presented techniques that help you catch bad input and give the user another chance to input good data. In addition, printing to a file rather than to paper can help save trees. This lab provides practice in the skills needed to build robust applications.

Chapter Highlights

- An exception is an event occurring during the execution of an application that disturbs the flow of instructions. Examples are division by zero, data type mismatch of input, or a String index out of bounds.
- As an alternative to throwing exceptions, programmers can use a try/catch block to continue the processing, if possible.
- An exception handler can catch an exception and provide instructions to avoid a crash.
- The try/catch/finally block can be used to catch both checked and unchecked exceptions.
- Input validation ensures user input matches expectations.
- Use a loop to accept input, perform validation, and ask again if input is improper.
- The PrintWriter class provides methods to identify the destination path, determine if a new file is needed or a file already exists, create output, print to the file, and close it.

While studying, look for the activity icon for:
- Vocabulary terms with e-flash cards and matching activities.
- Starter files for lab activities.

These activities can be accessed at
www.g-wlearning.com/informationtechnology/1773

Warm-Up Exercises

_____ 1. Consider the following code segment. Which exception is thrown if the user enters 42.0?

```
Scanner input = new Scanner( System.in );
System.out.print( "Enter Jackie Robinson's jersey number: " );
int jerseyNumber = input.nextInt( );
```

A. InputMismatchException
B. OutOfBoundsException
C. DivisionByZero
D. DataTypeMismatch
E. No exception is thrown.

_____ 2. Consider the following code segment. What is the limitation in this sequence?

```
Scanner input = new Scanner( System.in );
String s = "";

System.out.print( "Enter an integer: " );
try {
   int integer = input.nextInt( );
   s = String.format( "The integer is %d.", integer );
}
catch ( InputMismatchException e ) {

   s = "There is a problem with your entry. It was not an integer.";
}
finally {

   System.out.println( s );
} // end try/catch/finally
```

A. An exception is thrown no matter what the user enters.
B. A loop is needed in case an exception gets thrown.
C. The exception object e is not printed.
D. Scanner should be used for output.
E. The variable identifier integer is a keyword.

_____ 3. Consider the following code segment. What is the limitation in this sequence?

```
final int NUMBER_OF_DIGITS = 16;
do {
   System.out.println( "Enter your credit card number; " + NUMBER_OF_DIGITS +
           " digits only" );
   inputCard = input.nextLine( );
   if ( inputCard.length( ) != NUMBER_OF_DIGITS ) {
      System.out.println( "Please enter " + NUMBER_OF_DIGITS
               + " digits!" );
   }
} while ( inputCard.length( ) != NUMBER_OF_DIGITS );
```

A. Validates both length and digits entered.
B. Does not validate digits input.
C. Ensures 16 digits are entered.
D. Prompts the user to encourage correct input.
E. Exception thrown if more than 16 digits are entered.

_____ 4. Consider the following code segment. The file f:\\FirstFile.txt contains "Hello World". What is the output?

```
try {
   // open the appropriate file
   File inputFile = new File( "f:\\FirstFile.txt" );
   Scanner file = new Scanner( inputFile );

   if ( file.hasNext( ) ) {
      String line1 = file.nextLine( );
      System.out.println( line1 );
   } else {
      System.out.println( "The file has no data. We are sorry for the omission." );
   }
} catch ( IOException e ) {
   System.err.println( "Caught IOException: " + e.getMessage( ) );
} // end try/catch on read file
```

   A. The file has no data. We are sorry for the omission.
   B. Caught IOException. File has no data.
   C. Hello World
   D. FileNotFoundException
   E. Caught IOException: unable to find file.

_____ 5. Select the method used to define a new file for output.
   A. Scanner output = new Scanner(system.out);
   B. PrintWriter output = new FileWriter("Output.txt ")
   C. FileWriter output = new FileWriter ("Output.txt", false);
   D. FileWriter output = new FileWriter ("Output.txt", true);
   E. Scanner output - new Scanner (system.out, "Output.txt");

_____ 6. Which of the following sends an exception object to the runtime system?
   A. catching an exception
   B. checked exception
   C. exception handler
   D. throwing an exception
   E. unchecked exception

_____ 7. Which of the following is the process of ensuring the information entered into the program is acceptable?
   A. input validation
   B. user experience
   C. exception processing
   D. data typing
   E. scanning

_____ 8. Which of the following is always processed in a try/catch/finally block?
   A. try/catch/finally
   B. try only
   C. try/catch
   D. try/finally
   E. finally only

_____ 9. The _____ method releases resources allocated to a file.
   A. print( )
   B. close( )
   C. println( )
   D. output( )
   E. nextLine( )

_____ 10. Which of the following is the process of adding data to the end of a file?
   A. output validation
   B. overwrite
   C. append
   D. upend
   E. scan

Lab 11-1

Handling Exceptions

Programmers writing applications that accept input from a user or an external file face concerns with expected cooperation. Once the code is written and deployed, it is up to the user to provide valid input either live or from a file. The programmer can provide exceptionally good prompts for the user. In the end, however, the user must cooperate and provide valid input. The only recourse for the programmer to keep the code from crashing is to anticipate typical user errors and handle any exceptions to keep the program running.

In earlier chapters, a graceful exit from a program was assisted by throwing exceptions when a problem arose. In this lab, you will use other means to keep the program running while handling erroneous situations. Before beginning this activity, download the files for this lab from the student companion website.

Learning Goals
- Identify possible user errors.
- Use a try/catch/finally block to handle exceptions.

Materials
- jGRASP Integrated Development Environment
- Starter files from the student companion website

Name _____

Application and Extension of Knowledge

Validating Input for a Division-Practice Game

Arithmetic exceptions are checked in Java programs. An exception can be thrown to catch an exceptional arithmetic condition. When Juan's little brother needed practice with his division facts, particularly quotient and remainder, Juan wrote a program to randomly generate 10 division problems. With one exception, Juan did a fine job. In this activity, you will write a program that detects and corrects the error Juan made.

Procedure

1. Launch jGRASP, and open the DivisionPracticeSnippet.java file. Examine the code to see how it will execute.

2. Compile and run the application. Use the following test input:

    dividend = 50

    divisor = 5

    Describe the output in the space below.

    _____

    _____

    _____

    _____

    _____

3. Name the kind of error that occurred: compiler, logic, user, or runtime. Explain how to fix it.

    _____

    _____

    _____

4. Locate comment 1. Edit the line that generates a random divisor to match this:

    ```
    int divisor = rand.nextInt( upperDivisor ) + 1;
    ```

    Compile and run again. Describe the output in the space below.

    _____

    _____

    _____

Reflections

1. The exception could have been handled with a try/catch/finally block. Explain why you would or would not use that code in this situation.

    _____

    _____

    _____

2. Why did you not have to fix the line that generates a random dividend? Explain why there is never an exception when the dividend is zero, but the divisor is not zero.

___

___

___

Reading from a File for a Number Game

Many parents play number games with their children. This helps children build their number facts and mental computation skills. Practically anything could be used for a number game. In this activity, you will write a program to read a file containing a counting rhyme, find the integers, and add them. Before you begin, make sure the SumNumbersSnippet.java and tokens.txt files are saved in the same folder.

Procedure

1. A useful method for reading integers in a file is the hasNextInt( ) method for the Scanner class. Apply what you learned about researching the Java Class Library to find the API for the hasNextInt( ) method. Write the API in the space below.

___

___

___

___

2. Launch jGRASP, and open the SumNumbersSnippet.java file. Locate comment 1. Add the import for the two packages java.io.* and java.util.*. Explain why the asterisks are required in the import statements.

___

___

___

___

3. Notice you are not throwing an exception because you are going to use a try/catch block. Describe the difference in actions using these two approaches.

___

___

___

___

4. Locate comment 2. To accommodate exception handling, you need a try block. What parts of the code need to be inside the try block? Explain in the space below.

___

___

___

___

5. Locate comment 3. The code in this section has been completed for you using the hasNextInt( ) method. Describe what is happening in the if statement.

   _____
   _____
   _____
   _____

6. Locate comment 5. Write the code to display the sum.
7. Locate comment 6. What kinds of exceptions are caught when the program throws an IOException? List at least two exceptions in the space below that might occur in the current program.

   _____
   _____
   _____
   _____

8. Compile and run the program with the downloaded tokens.txt file.
9. Locate another counting rhyme and create your own tokens.txt. Run the program with the new file. Do you need to recompile? Why or why not?

   _____
   _____
   _____
   _____

Reflections

1. Explain why this program requires a try/catch block. Did it avert termination of the program in exception conditions?

   _____
   _____
   _____
   _____

2. Consider the possibility the user might want to see the token file. It would be relatively easy to read in the lines and print them out before entering the search loop. However, after reading the file, the program is at the end of the file. Think of a way to start at the beginning of the file again.

   _____
   _____
   _____
   _____

Validating Input of Money

You can provide precise instruction to the user for input, but the user may not comply. In this lab, you will attempt to accept a dollar amount from a user. Users may use the dollar sign. You will scan in a String for the whole input line. If there is a dollar sign, you will ignore it. Then, you will convert the String to a double using the Double wrapper class.

Procedure

1. Launch jGRASP, and open the InputMoneySnippet.java file. Compile and run it twice, once with a dollar sign and once without. Describe the output.

   _____

   _____

2. Redesign this program to try to minimize user input errors. Change the input to a String. Apply the String method trim( ) to remove extraneous leading and terminating spaces.

   Edit

   ```
   // get input of double value
   double dCost = input.nextDouble( );
   ```

   to

   ```
   // get input as a String and remove leading and trailing spaces
   String sCost = input.nextLine( );
   sCost = sCost.trim( );
   ```

   Is the program ready for running? Has this handled the extra $?

   _____

   _____

3. Handle the dollar sign error by checking the first character of sCost for a $. Which String manipulation methods will be useful here?

   _____

4. Now you have a String that you hope has a double value in it. There is a useful Double wrapper class method to change the String value into a double value. Apply what you learned about researching the Java Class Library to find the API for the Double.parseDouble( ) method. Write the API in the space below.

   _____

   _____

5. Add the code to convert the String sCost into the double dCost. Compile and run the program. Use test data 15, 15.00, $15, 15 with trailing and leading spaces, and $15 each. Describe the output.

   _____

   _____

6. It appears you need to insert a try/catch block. Use the space below to plan. Your goal is to gracefully exit the program and not crash as a result of an exception.

Reflections

1. Consider other types of errors users might make in data entry for this program. Explain which ones would generate the exception.

2. Consider the final code. At what line does the exception actually occur?

Debugging Challenge

Consider the following code segment. What is the reason for the compiler error listed below? Record your response in the space provided.

```java
try {
   dCost = Double.parseDouble( sCost);
   s = String.format( "The cost of the tickets is $%.2f.", dCost );
   System.out.println( s );
}
catch ( NumberFormatException ) {
   System.err.println( "Caught NumberFormatException: " + e.getMessage( ) );
   System.out.println( "Numeric values for the ticket price were expected." );
   System.out.println( "Program terminating." );
}
```

Error message:

```
InputMoneySnippetAnswer.java:30: error: <identifier> expected
   catch ( NumberFormatException ) {
                               ^
```

Lab 11-2
Data Validation and Output

Because good answers require good input, data validation is an important part of programming. Users need all the help you can give them to understand how to format their input. However, mistakes can and will happen. It is the job of the code to catch these errors and try to keep the program running.

Saving information to a file makes it easier to store the results and share them with other people and programs. This lab will help you create files and save the output that otherwise would be shown on the computer display. Before beginning this activity, download the files for this lab from the student companion website.

Learning Goals
- Validate input from user.
- Get current year from the Calendar class.
- Write information to a file.

Materials
- jGRASP Integrated Development Environment
- Starter files from the student companion website

Application and Extension of Knowledge

Capturing and Validating a Birth Date

Think about the apps you have used that have asked for your birth date. You have probably noticed that apps use many different formats for entering a date. Some apps request the format of mm/dd/yy, while others request the format mm/dd/yyyy. Some provide scrolling lists to capture the name of the month, day, and year. Entering each part of the date separately often minimizes input errors, but data validation must occur, nevertheless. Although handling an exception and gracefully exiting is useful, these actions are reserved for fatal flaws. At all times, it is best to keep the program running.

In this activity, you will create a program that prompts the user for a date and validates the input. Think about all the input errors a user might make. Accommodate these user errors in the code.

Procedure

1. Launch jGRASP, and open the InputBirthdateSnippet.java file. Compile and run the program, following the format in the prompt exactly. Make notes in the space below about what you experienced.

2. The app expects a user to follow the format in the prompt (mm/dd/yyyy), including the slashes. Look at the code to find Integer.parseInt( ). This is an analogous use of the Double wrapper class method Double.parseDouble( ) used in the previous lab. This assumes the user complies with the input rules. If the format is followed, the app asks the user if the input is correct. This is called *user verification.* Run the app again and enter the date 5/3/2000. Describe the outcome in the space below.

3. The NumberFormatException occurs because the first two digits of the date were read as numeric and a slash is not numeric. The first task is to handle the exception with a Scanner feature. A Scanner can use a String as input and use a delimiter, such as a slash. Apply what you have learned about researching the Java Class Library to find the API for Scanner contructors and methods. Locate the constructor that uses a String for its argument. Locate the useDelimiter( ) method. Write the APIs in the space below.

4. Modify the birth date input by inserting this after getting the String birthdate:

```
Scanner line = new Scanner( birthdate );
line.useDelimiter( "/" );
```

Then edit the mm, dd, yyyy conversions to:

```
// parse the input string, skipping over the /'s
int mm = line.nextInt( );
int dd = line.nextInt( );
int yyyy = line.nextInt( );
```

Compile and run the program. Try the input 5/3/2000 again. Describe the result in the space below.

5. Suppose the user enters the date in a different format than requested. Run the program, and enter the date May 3, 2000. The program terminates in an InputMismatchException. You could throw an InputMismatchException to provide a clean exit from the program, but ideally the program should keep running so the user can enter the date in the correct format. The loop is there to provide another chance to follow the format. Insert a try/catch block inside the loop to catch the exceptions and keep running. This is another attempt to encourage the user to use the provided format. Ultimately, it is up to the user to comply. Compile the program and run it using a variety of valid dates.

6. Once the NumberFormatException is avoided, a date such as 14/45/2120 can be entered. However, that is not a valid birth date, for certain. You can write validation code for each part of the date to show the user what the input error is and prompt to try again. Locate comment 1. Insert code to validate a user has entered a proper number for the month. If not, set the variable validDate to false. Plan your work below. Then, compile and run the program, testing with 14/45/2120 and your own birth date.

_____

_____

_____

_____

_____

_____

7. Locate comment 2. Validate the day. This will take some planning. In the space below, list all months and the number of days in the month. Look for a pattern. Think about using a switch statement with multiple cases performing the same code. Insert code to validate a user has entered a proper number for the month. If not, set the variable validDate to false. Then, test with 11/18/2120 and your own birth date.

_____

_____

_____

8. There are two items to validate for the year: 1) that it is four digits, and 2) that it is before this year. Locate comment 3. Think of a way to validate that yyyy contains four digits. You might consider looking at the original input of the String birthday. Plan your work in the space below. Then, test with 12/18/2120 and your own birth date.

_____

_____

_____

9. Locate comment 4. Know your user. It is not likely that an infant who is less than one year old is using the app. It would be easy to hardcode the current year, but if your app is to have longevity, you must be able to detect the current year in which the app is running. Fortunately, the Java Class Library has a Calendar class. Apply what you have learned about researching APIs to find the Calendar class and how to get the value for the current YEAR. Use the space below for taking notes from the API. Import the class, write the instantiation statement and the method to use to get the YEAR. Test with 12/18/2002 and your own birth date.

_____

_____

_____

_____

_____

Reflections

1. Describe the test cases you would use to fully test this app.

2. What do you think could be done to validate that the date entered actually is the birth date of the user?

Reading from a File to Identify Near-Earth Asteroids

NASA maintains a large database of asteroids. The SelectedAsteroidsNasa.txt file contains a list of asteroids of interest to the general public. Some of these, but not all, are in a near-earth orbit (NEO). In this activity, you will open and read the SelectedAsteroidsNASA.txt text file, search for the string "Near-Earth" in every line of the file, and write those records to a new file named NEO.txt. Before you begin, make sure the SearchAsteroidsSnippet.java and SelectedAsteroidsNasa.txt files are saved in the same folder.

Procedure

1. Apply what you learned about reading from and writing to a file. Using the following algorithm in steps 1–9. Enter the API for the methods you will need.

    ```
    /***** 1. Open the SelectedAsteroidsNASA.txt file for input.
    */
    ```

    ```
    /***** 2. Create a new NEO.txt file to write only.
    */
    ```

```
/***** 3. Instantiate the Printwriter object.
*/
```

```
/***** 4. Define a boolean flag, stringFound as false.
*/
```

```
/***** 5. Begin the while loop to read the file line by line.
*/
```

```
/***** 6. Read a line, and search for the search string.
*/
```

```
/***** 7. If the search string is found,
            write the line to the output file, and the display,
            and set stringFound to true.
*/
```

```
/***** 8. Check if the search string was found in any line.
            Output a message if no match was found.
*/
```

```
/***** 9. Close the NEO.txt file.
*/
```

2. Launch jGRASP, and open the SearchAsteroidsSnippet.java file. Write the code using the methods you identified. Compile and run. Enter a summary of the output below. Verify that the NEO.txt file contains the same output.

Reflections

1. Suppose the file SelectedAsteroidsNASA.txt were replaced by a file containing the entire data set for asteroids. Explain how the code would need to change.

2. Open the SelectedAsteroidsNASA.txt file. Explore its structure. Explain how you would parse the input line to save the asteroid number, its name, and the notes. What data types would you use? What input statements would you use?

Debugging Challenge

This program intends to keep a history of books read each year. It captures the number read and appends a line to the file **BooksRead.txt**.

```
/* Debugging Challenge Chapter 11*/
import java.io.*;
import java.util.Scanner; // this imports the Scanner class
import java.util.Calendar; //  this imports the Calendar class to get the year.

public class Debug11_2 {
   public static void main( String [ ] args ) throws IOException {
      Scanner input = new Scanner( System.in ); // create the input object

      /*Prompt the user for the number of books they have read
         this year, read the value into a variable of the appropriate
         type, and append the value in a message to a file
      */
      System.out.print( "How many books have you read this year? " );
      int booksRead = input.nextInt( );

      // Open BooksRead.txt for appending.
      FileWriter fileWriter = new FileWriter( "BooksRead.txt", false );
      PrintWriter printWriter = new  PrintWriter( fileWriter );
      Calendar cal = Calendar.getInstance( );

      String s = "In " + cal.get( Calendar.YEAR ) + " I read " + booksRead;

      printWriter.println( s );
      printWriter.close( );

   } //end main
} // end class
```

The program compiles without errors, but after the program runs twice, the file has only one line containing the number of books the second user entered along with the current year. What is wrong, and how can you fix it? Record your response in the space provided.

Name _____ Date _____ Class _____

CHAPTER 12
Custom Classes and Methods

Often in app development, coders are presented with multiple entities that behave in a similar manner and are acted on by the program in the same way. Defining classes and creating instances of those classes protects the properties of the instances and supplies methods for manipulating those instances, thereby simplifying the code. You can learn much about custom classes by applying what you already learned about the Java Class Library. In addition, think about the needs of the application in manipulating these objects. Doing so will provide insight into types of properties that should be assigned and methods that are required to manage those properties.

Chapter Highlights

- A class consists of data and methods, and each object created from a class has its own copy of the data stored in the class's instance variables, but all objects of a class share the class methods.
- A class should provide a constructor for the client to be able to create objects; otherwise, the compiler's default constructor is used, which assigns default values to the instance variables based on the data type of the variable.
- An accessor, which is a value-returning method, allows the client to view data in an object and follows the basic pattern of having a name starting with get followed by the name of the instance variable, not taking parameters, and returning the same data type as the instance variable.
- A mutator method allows a client to change the values of the data in an object and follows the basic pattern of having a name starting with set followed by the name of the instance variable, taking one parameter that is the new value for the instance variable, and usually is a void method.
- Preceding a name with this refers to an instance variable, and the keyword is used to prevent hiding the instance variable if a parameter has the same name as the variable.
- A custom class should include a method that completes the intended function of the class, and it should also include a toString method to return the data of the class as a String.

While studying, look for the activity icon for:
- Vocabulary terms with e-flash cards and matching activities.
- Starter files for lab activities.

These activities can be accessed at
www.g-wlearning.com/informationtechnology/1773

- A subclass inherits public methods from a superclass; inheriting from a superclass is defined as an "is-a" relationship because each object of the subclass "is" also "an" object of the superclass.
- Place the code for constructing a graphical class in a separate method from the constructor to clean up the constructor code; the shapes composing the figure should be drawn relative to an origin (0,0) for the figure, which is used by the JavaFX runtime system to place the figure in the graphics window.

Warm-Up Exercises

For questions 1–9, consider the following class definition.

```
public class Pet {
   private String type;
   private double weight;
   private String sound;

   public Pet( ) {
   }

   public Pet( String  ty, double wt, String sd ) {
      type = ty;
      weight = wt;
      sound = sd;
   }
} // end class
```

_____ 1. Which of the following are the instance variables of the class?
   A. Pet, type, weight, sound
   B. type, weight, sound
   C. ty, wt, sd
   D. Pet, ty, wt, sd
   E. Dog, 20, 0, bark

_____ 2. A client program instantiates a Pet object using the following statement. What will be the value of the fido type?

   `Pet fido = new Pet( ) ;`

   A. null
   B. 0.0
   C. false
   D. 0
   E. Unable to determine.

_____ 3. A client program instantiates a Pet object using the following statement. What will be the value of the fido weight?

   `Pet fido = new Pet( ) ;`

   A. null
   B. 0.0
   C. false
   D. 0
   E. Unable to determine.

_____ 4. A client program instantiates a Pet object using the following statement. What will be the value of the Pet weight?

```
Pet lizard = new Pet( "anole", 1.2, "No sound" ) ;
```

A. lizard
B. 1.2
C. No sound
D. anole
E. Unable to determine.

_____ 5. Which of the following is the correct code for an accessor method for the Pet class?

A.
```
private String getWeight( ) {
    return weight;
}
```

B.
```
public String getWeight( ) {
    return weight;
}
```

C.
```
private double getWeight( ) {
    return weight;
}
```

D.
```
public double getWeight( ) {
    return weight;
}
```

E.
```
public int getWeight( ) {
    return weight;
}
```

_____ 6. What is the correct code to write in the body of this mutator method for the Pet class?

```
public void setSound( String sound ) {
    //   ??
}
```

A. return sound = this.sound;
B. return this.sound = sound;
C. this.sound = sound;
D. sound = this.sound;
E. sound = sound;

_____ 7. Which of the following headers would be correct for the toString method for this class?

A. private void toString( )
B. public void toString(String type, double weight, String sound)
C. public String toString(String type, double weight, String sound)
D. public String toString( )
E. private String toString( )

_____ 8. Which of the following statements would be correct code for the body of the toString method?

A. System.out.println("The " + type + " says " + sound + ".");
B. return "The " + type + " says " + sound + ".";
C. System.out.println(return ("The " + type + " says " + sound + "."));
D. return "The dog says bark.";
E. String toString =   "The " + type + " says " + sound + "." ;

_____ 9. What code would be correct for a method that adds pounds to a pet's weight?

A.
```
public void addWeight( ) {
    weight += pounds;
}
```

B.
```
public void addWeight( double pounds ) {
    weight += pounds;
}
```

C.
```
public double addWeight( double pounds ) {
    weight += pounds;
}
```

D.
```
public double addWeight( double pounds ) {
    return weight + pounds;
}
```

E.
```
public String addWeight( String pounds ) {
    return "weight + pounds";
}
```

_____ 10. The class definition for a graphical class begins with the following statements. Which of the following would be a correct statement to put in the constructor?

```
public class MySprite extends Sprite {
    public MySprite( Group root, double x, double y ) {
```

A. Sprite(root, x, y)
B. MySprite(root, x, y):
C. Super(root, x, y);
D. super(root, x, y);
E. super.MySprite(root, x, y);

Lab 12-1

Creating Classes

In this lab, you will begin writing a game called Throw Down in which two players compete by throwing two dice. Depending on whether the total roll is doubles, odd, or even, points will be added or subtracted from the player's score. A winning score is 100 and a losing score is 0. Because two players will perform the same activities and earn points in the same way, it makes sense to create a Player class for this program. Before beginning this activity, download the files for this lab from the student companion website. The Die class and Player class files should be in the same folder.

Learning Goals
- Define a class.
- Create default and overloaded constructors.

Materials
- jGRASP Integrated Development Environment
- Starter files from the student companion website

Application and Extension of Knowledge

Creating a Player Class

In this activity, you will begin creating the Player class and the game client. With this class, you will be able to create instances of individual player objects and easily manage their properties. This makes game-application code more streamlined, simpler, and easier to follow. You will also reuse the Die class you created for Chapter 12 of the text to manage the roll of two dice. Be sure that you also downloaded the Die class from the student companion website.

A Player object will need a name and will need to be able to roll the dice and to keep track of its score. This leads to the design decision that the instance variables of the Player class should be a name, a score, and two dice. The Die class created in the textbook can be reused here.

For this class, only one constructor will be provided, which accepts initial values for the name and score. By omitting the default constructor, the client of the class is forced to assign a name and score to each Player object as it is instantiated.

Procedure

1. Follow these instructions and plan your solution in the spaces provided.

    ```
    /***** 1. Define the instance variables:
            name should be a String
            score should be an int
            die1 and die2 should be objects of the Die class
            All instance variables should be private.
    */
    ```

    ```
    /***** 2. Define the constructor.
            The constructor should accept a name and a beginning score
            as parameters and store those values in the instance variables.
            In addition, it should roll the two dice.
    */
    ```

2. Launch jGRASP, and open the Player.java starter file. You will see it contains only the class definition line and the open and closing braces.
3. Enter the code you wrote above.
4. Compile the code and fix any errors.

5. Open the ThrowDownSnippet.java class file. This will be the client of the **Player** class and will manage the playing of the game. At this point, you will simply define point values for the game and instantiate two **Player** objects.

6. Follow these instructions and plan your code in the spaces provided.

```
/***** 1. Define constants for point values:
            an even roll earns 10 points
            an odd roll subtracts 5 points
            doubles earns 20 points
            winning score is 100
*/
```

_____
_____
_____
_____
_____
_____

```
/***** 2. Instantiate two Player objects. Assign names of your choosing,
            and set the beginning score to 10 points.
*/
```

_____
_____
_____

```
/***** 3. Output a greeting that the game is about to begin. Follow this with two
            statements that consist of a tab and the object reference for one of the
            Player objects. This will simply output the hash code at this time, but it will
            prove the Player objects have been created.
*/
```

_____
_____
_____

7. Enter the code you wrote above.
8. Compile the code and fix any errors.
9. Run the code and verify that the two **Player** objects have been created.
10. Keep these two files. In the next lab, you will complete the **Player** class and the Throw Down game.

Reflections

1. If you did write a default constructor, what values would you assign to the instance variables? Name one problem that would arise with assigning a default name to a Player.

2. Reuse of prewritten classes is an advantage of object-oriented programming. Explain how reuse is implemented in this program and why it is an advantage.

Debugging Challenge

In your program, you write this code to define a class:

```
public class MyClass {
   private int number;
   public MyClass( int number ) {
      this.number = number;
   }
}
```

You write this client code to instantiate an object of the MyClass class:

```
public class MyClassClient {
   public static void main( String [ ] args ) {
      MyClass mc1 = new MyClass( );
   }
}
```

The class compiles without errors, but when you compile the client code, you get this error:

```
error: constructor MyClass in class MyClass cannot be applied to given types;
```

What is wrong, and how can you fix the code? Record your response in the space provided.

Lab 12-2
Defining Methods

Consider what actions and reactions the Throw Down player will have in the game. Manage properties and build methods to accommodate these. In this lab, you will complete the Player class and the Throw Down game. Before beginning this activity, download the files for this lab from the student companion website.

Learning Goals
- Define a value-returning method.
- Construct accessor and mutator methods.
- Use the this keyword.
- Write a toString method.

Materials
- jGRASP Integrated Development Environment

Application and Extension of Knowledge

Writing Methods to Complete a Class

In this activity, you will complete the Player class and the Throw Down game you started in Lab 12-1. In the previous lab, you defined the Player instance variables and a constructor. The class will need the usual accessors and mutators and a toString method. In addition, the class will need one more method (play) that rolls the dice for the Player turn. You will write methods to handle rolling dice and determining the winner/loser. Specifically, the class will need these methods:

- Accessors: getName, getScore, getDie1Roll, getDie2Roll
- Mutators: setName, addPoints, subtractPoints
- Others: toString, play

Procedure

1. Start with the Player.java file completed in Lab 12-1. Comments 1 and 2 were completed in that lab, so you will begin with comment 3. Plan your solution in the space provided.

   ```
   /***** 3. Write the accessor and mutator for name.
   */
   ```

```
/***** 4. Write the play method. This method should roll die1 and die2
         and return the combined roll.
 */
```

```
/***** 5. Write the accessors for the roll of the two Die objects.
         These methods should call the getRoll method of the Die class.
 */
```

```
/***** 6. Write the accessor for the score. Also write the mutators: addPoints
         and subtractPoints.
 */
```

```
/***** 7. Write the toString method. It should return the
         name and score.
 */
```

2. Launch jGRASP, and open Player.java file created in Lab 12-1.
3. Enter the code you wrote in question one.
4. Compile that code and correct any errors.

Consider the tasks the ThrowDownSnippet must accomplish to manage the playing of the game. The game must alternate turns between the two players. Given that each player will perform the same functions (rolling the dice) and each player's score will be identically altered depending on the value of the roll, it makes sense to write that code just once. The only thing that will be different in the code is which player is taking the turn. This can work by creating a Player object reference that will be used to point to the current player. Then, alternate assigning the players to this reference and use the reference throughout the turn. This reference can be defined as:

```
Player current;
```

Once the correct Player object has been assigned to current, call the play method. This method returns the sum of the two dice rolls. Next, check if the roll consists of doubles, then if the roll is even or odd, and finally add or subtract points accordingly. All this should take place in a loop that checks for whether or not the current player has reached a winning score or has dropped to 0. Here is the output from a sample run of the program. Of course, the dice rolls will likely be different each time you run the game.

```
Beginning the game! The players are:
   Jordan has 10 points.
   Adrian has 10 points.
Jordan's turn.
   The roll is 6
   Roll is even! Add 10 points.
   Jordan has 20 points.
Adrian's turn.
   The roll is 5
   Roll is odd! Subtract 5 points.
   Adrian has 5 points.
Jordan's turn.
   The roll is 8
   Doubles! Add 20 points.
   Jordan has 40 points.
Adrian's turn.
   The roll is 7
   Roll is odd! Subtract 5 points.
   Adrian has 0 points.

Sorry! Adrian loses.
```

5. Open the ThrowDownSnippet.java class created in Lab 12-1. This is the client of the Player class and manages the playing of the game. At this point, you have completed comments 1, 2, and 3; that is, you have defined point values for the game and instantiated two Player objects. Start with comment 4. Plan your solution in the spaces provided.

```
/***** 4. Define a Player reference named current to use for the player whose turn it is.
*/
```

_____

_____

_____

_____

_____

/***** 5. Start a while loop to continue play until the current player (current)
          has either won or lost.
          Winning means the score is at least WINNING_SCORE;
          losing means that the score is 0 or below.
          Applying De Morgan's Laws can help you code the correct condition.
          Also, code the open and closing braces now so you do not need
          to remember to do so later.
*/

/***** 6. Inside the while loop, code an if statement to alternate turns. Assign current
          to the player that is not now assigned.
*/

/***** 7. Announce the current Player name and call the play method.
          Output the roll.
*/

```
/***** 8. Check for doubles, then check for even, or odd,
          and add or subtract points accordingly.
          Call methods of the Player class as needed.
          Output a message saying what rule applies
          and how the scores are affected.
          Finally, output the current player's name and new score.
  */
```

```
/***** 9. After the while loop ends,
          check to see whether or not the current player has won or lost.
          Output a message with the player's name and final score.
  */
```

6. This completes the game.
7. Compile and run the Throw Down code.
8. Test this code with the Player class. You may need to run the code several times to verify the correct number of points is being added or subtracted from the scores and that the correct winner is reported.

Reflections

1. Why do you think the game starts the players with 10 points rather than 0?

2. Why does the check for doubles need to take place before checking for even or odd rolls?

Debugging Challenge

Consider this code, which is intended to be a mutator for a class that has a **direction** instance variable. The instance variable is an **int**. Valid values for direction are between 0 and 359, inclusive. Here is the class:

```
public class MyClass {
   private int direction;

   public MyClass( ) {
      this.direction = 0;
   }
   public void setDirection( int direction ) {
      if ( direction >= 0 && direction <= 359 ) {
         direction = direction;
      }
   }
   public String toString( ) {
      return "Direction is " + direction;
   }
}
```

This is the code in main in the client:

```
MyClass mc1 = new MyClass( );
System.out.println( mc1.toString( ) );
mc1.setDirection( 300 );
System.out.println( mc1.toString( ) );
```

The code compiles without error, but when you run the code, the output is:

```
Direction is 0
Direction is 0
```

As you can see, **direction** is still 0 after the mutator is called with a valid value. What is wrong, and how can you fix the code? Record your response in the space provided.

Lab 12-3

Creating a Graphical Class

In this lab, you will create a Ball class that inherits from the Sprite class. You will also write several methods for the Ball class and call the methods from a client class. Before beginning this activity, download the files for this lab from the student companion website.

Learning Goals
- Use inheritance to create a graphical subclass.
- Write methods for a graphical class.
- Gain more practice in writing while loops.

Materials
- jGRASP Integrated Development Environment
- Starter files from the student companion website

Application and Extension of Knowledge

Constructing a Ball Class

In this activity, you will define a Ball class and write two methods that animate the ball. The client will draw a box on the window. The first method of the Ball class bounces the ball in the box. The second method of the Ball class moves the ball around the box, ricocheting against the walls of the box.

The creation of the Ball should be straightforward given the chapter code. The Ball should extend (inherit from) the Sprite class. Creating the ball consists of defining a Circle shape and specifying its center, radius, and color and then adding the Circle to the Group.

Remember, because the Ball inherits from the Sprite class, a Ball "is-a" Sprite. That means methods of the Ball class can call any method of the Sprite class without using an object reference. For example, to move the Ball forward 100 pixels, a method in the Ball class could use this code:

```
forward(100);
```

The first method bounces the Ball and has this API:

```
public void bounce( double boxBottom )
```

where boxBottom is the *y* location of the bottom of the box drawn on the window.

To bounce the ball, you should "drop" the ball from its current position by moving to the bottom of the box. Then, move the ball halfway up to its starting *y* location. Repeat this until the distance between the return position and the bottom of the box is 0. A while loop should serve well. Here are some considerations:

- Because the ball returns up only halfway from its starting point for each bounce, you need to keep track of the starting position for each bounce iteration.
- Remember that *y* values get larger as you move down the window.
- The (*x*, *y*) location of the ball is the center of the Ball object. For the lower edge of the ball to bounce off the box bottom, the target *y* location for the bottom of the bounce should be boxBottom − radius. This means that the radius needs to be an instance variable of the Ball class.

Putting this together, the algorithm is:
1. Calculate the target location of the bottom of the box.
2. Save current *y* location as the start *y*.
3. While start *y* < target bottom:
   A. Face down.
   B. Move to the bottom.
   C. Face up.
   D. Move halfway to start *y*.
   E. Save the new start *y*, which is the loop update.

The second method ricochets the ball off the four walls of the box and has this API:

```
public void ricochet( double boxTop, double boxLeft, double boxBottom, double boxRight )
```

where boxTop is the *y* value for the top of the box, boxLeft is the *x* value for the left side of the box, boxBottom is the *y* value for the bottom of the box, and boxRight is the *x* location for the right side of the box. This method moves the ball to the top center of the box, then rolls the ball around the box in the directions shown in the figure below:

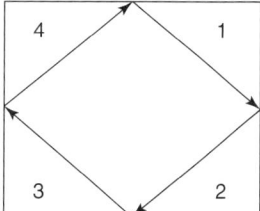

In other words, the ball starts at the top center, then moves diagonally down and to the right until it hits the right side of the box. Next, it moves diagonally down to the left until it hits the bottom of the box. Then, it moves diagonally up to the right until it hits the left side of the box. Finally, it moves diagonally up to the right until it hits the top of the box. This algorithm has some considerations as well:

- The ball needs to be moved to the top center of the box. This location can be calculated. The *x* value is half the distance between *x* values for the left and right walls. The *y* value is simply the top wall. These values can be sent to the moveTo method.
- The ball also needs to face diagonally down to the right. That heading is 135, but the direction the ball is currently facing is not known. To ensure the heading is 135, you can get the heading of the ball and then turn the ball right with this formula: (360 − currentHeading) + 135
- The ball needs to "roll" until it reaches the side wall. You can use a while loop. Inside the loop, move the ball forward 5 pixels with each iteration as long as the *x* value is less than the boxRight value.
- To orient the ball so it is ready to move down to the left, turn it 90 degrees to the right.

The remainder of the method consists of three more similar while loops, turning 90 degrees to the right after each loop finishes. You will create the Ball class first.

Procedure

1. Launch jGRASP, and open **Ball.java** starter file. You will see that it includes the import statements and the skeleton of the **Ball** class.
2. Plan your solution in the space provided.

```
/***** 1. Insert "extends Sprite" in the definition of the Ball class.
*/
```

```
/***** 2. Define the instance variable radius, which will be the radius of
            the ball; radius is a double.
*/
```

```
/***** 3. Write the constructor.
            The constructor accepts parameters for Group root,
            as well as three doubles: the x and y location and a radius for the ball.
            a. Call super and pass the parameters on to the Sprite constructor.
            b. Set the radius instance variable.
            c. Call the buildBall method. The method does not return a value and takes
               no arguments. You will write this method next.
*/
```

```
/***** 4. Write the buildBall method,
            The method does not return a value and takes no arguments.
            This method should draw the ball.
            The ball is a Circle with center x and y being 0
            and the radius is the value of the instance variable.
            Set a color of your choosing for the Ball, and
            add the ball to the Group.
*/
```

Name _____

3. Enter the code you wrote above.
4. To test the code, open the TurtleBallSnippet.java file, which is the client of the Ball class. You will see some code is already written. A Turtle object has been created for the turtle to draw a box. The box boundaries are defined as constants. These constants will be used later as parameters to the Ball methods. You will also see a statement has been written to instantiate a Ball object.
5. Compile the Ball.java and BallSnippet.java files.
6. Run the BallSnippet program. Remember, the Ball.java class does not have a main method. You should see the ball appear at the top center of the box.
7. Go back to the Ball.java class, and write the first method to bounce the ball. Plan your solution in the space provided.

```
/***** 6. Write the bounce method to "bounce" the ball against the bottom wall of a box.
          The location of the bottom wall is sent as a parameter.
          The method does not return a value.
          Lift the ball's tail, and set the speed to Ball.FAST.
          Then, your code should implement the algorithm given in the lab instructions.
*/
```

8. Enter the code you wrote above into the Ball.java file.
9. Compile the file.
10. To test the bounce method, go to the BallSnippet, and uncomment the line of code that calls the bounce method. Then, run the BallSnippet application to see the ball bounce.

11. Write the ricochet method. It takes four parameters and does not return a value. Follow the algorithm in the lab instructions, and plan your solution in the space provided.

   ```
   /***** 7. Write the ricochet method. It takes four parameters: the top,
            left, bottom, and right walls of the box. Use the defined constants.
            The method should not return a value.
            Lift the tail so that the ball does not leave a trail.
            Start at the top middle of the box, then
            move the ball around the box diagonally. When the ball hits a wall, turn 90
            degrees to the right and head for the next wall.
            Continue until the ball returns to the starting position.
   */
   ```

12. Enter the code you wrote in the previous questions into the Ball.java file, and compile it.
13. To test the ricochet method, go to the BallSnippet and uncomment the line of code that calls the method. Then, run the BallSnippet application to see the ball roll around the box. If the ball does not roll as you expect, try commenting out all the code except for the first while loop. When that while loop works, uncomment the code for the second while loop, and so on. In this way, you can isolate and test your code.

Reflections

1. Often, more than one algorithm can solve a problem. In the ricochet method, instead of using while loops to move forward five pixels at a time, can you think of a way to calculate the target (x, y) location for each wall? If that can be calculated, could you simply use four moveTo method calls? How would you change the code to use this algorithm?

2. Devise a new method for the Ball class. Perhaps the ball will pass back and forth across the box. Perhaps, you have another creative idea. Write the method in the Ball class, and call the method from the BallSnippet program.

Debugging Challenge

Consider this code, which is intended to create the shape for a class that inherits from the Sprite class:

```
public void buildSquare( ) {
   Rectangle square = new Rectangle( );
   square.setX( 0 );
   square.setY( 0 );
   square.setWidth( 100 );
   square.setHeight( 100 );
   square.setFill( Color.BLUE );
} // end buildSquare
```

The code compiles without error, but when you run the code, the square does not appear in the window. What is wrong, and how can you fix the code? Record your response in the space provided.

Name _____ Date _____ Class _____

CHAPTER 13
Working with Arrays

Two of the key components of problem solving are pattern recognition and abstraction. These two components help you identify patterns and generalize a solution using the patterns. Often in computing, more than one item of data can be treated the same way. Similarities among the data allow you to collect them in a single unit and operate on them at the same time in the same way. The Java feature used to manage similar items of data is an array. Arrays promote easily read and understood code. In each problem-solving task you encounter, think about arrays if operations on the data are similar.

Chapter Highlights

- An array is an ordered sequence of data of the same data type, called elements; arrays are objects, so arrays must be instantiated.
- Array elements are referenced by an index, which can be used to retrieve the element value or change it.
- A for loop is used to traverse an array with a loop condition that the index is less than the length of the array, and this can be used to fill an array, print an array, find the average of the array, and find the minimum and maximum values in the array.
- A bar graph can be created from an array using a for loop and Rectangle objects; the value of each element is used to draw the height of each bar in the graph.
- Each element in an array of objects is a reference to the object, not the object itself; creating an array of objects requires instantiating the array and then instantiating each object in the array.
- The search key is the value to find, and a sequential search processes the array in order; an unsorted array mush be searched in its entirety if the search key is not found, but can be exited early if the key is found.
- The sort method of the Arrays class can be used to return the values in an array in a sequential order; searching a sorted array is more efficient than sorting an unordered array.

While studying, look for the activity icon 📲 for:
- Vocabulary terms with e-flash cards and matching activities.
- Starter files for lab activities.

These activities can be accessed at
www.g-wlearning.com/informationtechnology/1773

Warm-Up Exercises

_____ 1. Which of the following is the correct way to define an array of integers with four elements?
   A. int a = new int[4];
   B. int [ ] a = new int[4];
   C. int[4] = new int[];
   D. int [ ] a = new int[3];
   E. int a = new int[3];

_____ 2. What is the output of this code?
```
boolean [ ] b = new boolean[10];
System.out.println( b[9] );
```
   A. 0
   B. 1
   C. true
   D. false
   E. Unable to determine.

_____ 3. What statement will access the element whose value is 'c' with the following array?
```
char [ ] chars = { 'a', 'b', 'c', 'd', 'e' };
```
   A. chars[3]
   B. chars(3)
   C. chars[2]
   D. chars(2)
   E. chars['c']

_____ 4. What is the output of this code?
```
double [ ] d = new double[100];
System.out.println( d.length );
```
   A. 100
   B. 99
   C. 101
   D. 0.0
   E. 0

_____ 5. What is the size of the aa array defined below?
```
int [ ] aa = { 100, 2, 35, 6, 10 };
```
   A. 100
   B. 10
   C. 5
   D. 4
   E. 6

6. What is the output of this code?

    ```
    int [ ] ab = { 1, 2, 3, 4, 5 };
    int s = 0;
    for ( int i = ab.length - 1;  i >= 0;   i-- ) {
       s += ab[ i ];
    }
    System.out.println( s );
    ```

    A. 15
    B. 1 2 3 4 5
    C. 5 4 3 2 1
    D. 5
    E. 14

7. What is the output of this code?

    ```
    String [ ] names = { "Seth", "Rosalie", "Jose", "Avram", "Amy" };
    for ( int i = 0; i < names.length; i++ ) {
       if ( names[ i ].length( ) == 4 ) {
          System.out.print( names [ i ] + " " );
       }
    }
    System.out.println( );
    ```

    A. Jose Rosalie Seth Avram Amy
    B. Seth Jose
    C. Seth
    D. Rosalie Amy
    E. Jose

8. What is the output of this code?

    ```
    String [ ] cities = new String[ 3 ];
    cities[1] = "Orlando";
    cities[2] = "Dallas";
    for ( int i = 0; i < cities.length; i++ ) {
       System.out.print( cities[i] + " " );
    }
    System.out.println( );
    ```

    A. Orlando Dallas
    B. Dallas Orlando
    C. Dallas Orlando null
    D. null Dallas Orlando
    E. null Orlando Dallas

9. What is the output of this code?

    ```
    double [ ] dd = { 3.4, 6.7, 9.0, 10.1 };
    for ( int i = dd.length - 1; i >= 0; i-- ) {
       System.out.print( dd[i] + " " );
    }
    System.out.println( );
    ```

    A. 3.4 6.7 9.0 10.1
    B. 10.1 9.0 6.7 3.4
    C. 10.1 9.0 6.7
    D. 3.4 6.7 9.0
    E. 29.2

_____ 10. What is the output of this code?
```
int [ ] aaa = { 17, 23, 13, 2, 30 };
int m = aaa[0];
for ( int i = 0; i < aaa.length; i++ ) {
   if  ( aaa[i] < m ) {
     m =  aaa[i];
   }
}
System.out.println( m );
```
A. 17
B. 23
C. 13
D. 2
E. 30

Lab 13-1

Creating Arrays

Arrays make it possible to store multiple values of the same data type in memory. This allows programs to access the data and perform varied processing on the data as needed. The user can specify what processing is desired at runtime. Before beginning this activity, download the files for this lab from the student companion website.

Learning Goals
- Create an array.
- Populate an array with values.
- Access array elements.

Materials
- jGRASP Integrated Development Environment
- Starter files from the student companion website

Application and Extension of Knowledge

Creating a Sentence Analyzer Class Using an Array

In this activity, you will create a class that allows a client to analyze a sentence. The methods of this class will allow the client to retrieve the number of words in the sentence, retrieve a specific word from the sentence, or analyze the type of sentence (interrogative, declarative, or exclamatory) based on the last character in the sentence. This activity will call on your knowledge of creating a class and writing methods as well as your knowledge of traversing a String using a for loop and searching for specific characters in a String.

Name _____   Chapter 13 Working with Arrays **187**

   A SentenceAnalyzer class will need to store the sentence as a String, the last character as a char, and the individual words in the sentence as an array of Strings. These three items will be the instance variables of the class.

Procedure

1. Follow these instructions and plan your solution in the space provided.

```
/***** 1. Define the instance variables:
            a. a String to hold the sentence
            b. an array named words that will hold the words in the sentence; you do
               not know yet what size the array will be
            c. a char to hold the end punctuation of the sentence
            Remember that all the instance variables should be private.
 */
```

```
/***** 2. Write the header of the constructor.
            The constructor takes one parameter, a String,
            named userSentence.
            Code the open and closing braces.
            Instructions 2a through 2d will guide you through writing
            the body of the constructor.
 */
```

```
/***** 2a. Trim the userSentence parameter to remove leading and trailing spaces.
             Store userSentence in the sentence instance variable.
 */
```

```
/***** 2b. Extract the last character and store it in the instance variable.
             Remove the last character from userSentence.
 */
```

```
/***** 2c. To determine the size of the words array, you need to know how many
          words are in the sentence. The number of spaces in the sentence will
          be the number of words. First, append a space to the end of the
          userSentence to make it easier to count the last word. Then,
          traverse the userSentence and count each space found. Assume only
          one space separates words. Store the count in a variable named wordCount.
*/
```

```
/***** 2d. Now that you know the number of words in the sentence,
          you can instantiate the words array to be wordCount size.
          Be careful not to redefine the words array; just instantiate it.
          Traverse the userSentence again to store each word in the words array.
          For each space you find, extract the word, store the word in the array,
          then remove the word and space from userSentence.
          Note: be sure to modify the parameter userSentence
          rather than the instance variable sentence.
          This code will complete the constructor.
*/
```

```
/***** 3. Write the getNumberOfWords method.
         It returns the number of words in the words array.
         Hint: the length of the array.
*/
```

```
/***** 4. Write the getSentence accessor method.
         It returns the sentence instance variable.
*/
```

```
/***** 5. Write the getWordAt method.
         This method takes an index as a parameter.
         If the parameter is valid, return the word
         at that index. Otherwise, return an empty String.
 */
```

```
/***** 6. Write the getLastCharacter accessor method.
         It returns the lastChar instance variable.
 */
```

```
/***** 7. Write the getSentenceType method.
         Based on the lastChar, the method
         returns the String declarative (.),
         interrogative (?),
         exclamatory (!),
         or Unknown sentence type.
 */
```

```
/***** 8. Write the toString method.
         This method simply returns the sentence
         instance variable.
 */
```

2. Launch jGRASP, and open the SentenceAnalyzer.java starter file. You will see it contains only the class definition line and the opening and closing braces.
3. Enter the code you wrote in the previous question.
4. Compile the code, and fix any errors.
5. Open the SentenceAnalyzerSnippet.java file. This will be the client of the SentenceAnalyzer class. This client prompts the user for a sentence, instantiates a SentenceAnalyzer object, and then calls all the methods.
6. Enter the code you wrote in the previous questions.
7. Compile the code, and fix any errors.
8. Run the code and verify all methods in the SentenceAnalyzer class have been correctly written.

Reflections

1. The program assumes words are separated by only one space. If the user enters multiple spaces between words, how would the word count be affected?

2. Why do you need to count the words before instantiating the words array?

Debugging Challenge

An application has this code in main:

```
int [ ] myArray = new int[3];
myArray[0] = 10;
myArray[1] = 2.4;
myArray[2] = true;
```

When this code compiles, two errors are reported:

```
error: incompatible types: possible lossy conversion from double to int
    myArray[1] = 2.4;
               ^
error: incompatible types: boolean cannot be converted to int
    myArray[2] = true;
               ^
```

What is wrong, and how can you fix it? Record your response in the space provided.

Name _____

Lab 13-2

Processing Arrays

The name of an array is a reference to the values of the array. Attempting to output the name of an array results in printing the hash code rather than the elements. Because the name of the array is merely a reference, any processing to be performed on each array element should be done in a for loop that accesses each element one at a time. The standard format of a for loop that accesses each element from the beginning of an array to the last element is:

```
for ( int i = 0; i < array.length; i++ ) {
   //  access array[ i ];
}
```

Before beginning this activity, download the files for this lab from the student companion website. Be sure the highTemps.txt and TemperatureAnalyzerSnippet.java files are saved in the same folder.

Learning Goals
- Process an array element-by-element.
- Identify the steps for processing an array of objects.

Materials
- jGRASP Integrated Development Environment
- Starter files from the student companion website

Application and Extension of Knowledge

Using an Array for a Temperature Analyzer

In this activity, you will read high temperatures for a location from a file into an array. The temperatures are stored in the file highTemps.txt, with one temperature per line. Each temperature is an integer. Once the values are stored in the array, the code will find the highest and lowest temperatures and the average temperature. Then, the user will be prompted for a target temperature, and the number of days the temperature was at least the target temperature is output.

Procedure

1. Plan your solution in the space provided.

    ```
    /***** 1. Define and instantiate the temps array to hold NUMBER_OF_DAYS ints.
    */
    ```

    _____
    _____
    _____
    _____
    _____
    _____

```
/***** 2. Using a try/catch block, open the highTemps.txt file
          and read the contents into the temps array.
          The file has one temperature per line.
   */
```

```
/***** 3. Find the index of the highest temperature in the array.
          Output the highest temperature and the number of
          the day on which the highest temperature occurred.
   */
```

```
/***** 4. Find the index of the lowest temperature in the array.
          Output the lowest temperature and the number of the
          day on which the lowest temperature occurred.
   */
```

```
/***** 5. Find the average temperature and output the average as
         a floating-point number.
*/
```

```
/***** 6. Prompt the user for a target temperature and count the number of
         days that were at least that temperature.
         Output that number.
*/
```

2. Launch jGRASP, and open the TemperatureAnalyzerSnippet.java file. You will see a Scanner object has been instantiated and a constant has been defined for the number of temperatures in the text file.
3. Enter the code you wrote above.
4. Compile that code and correct any errors.

Reflections

1. It is important to ensure any reference to an index in an array is valid. Otherwise, an ArrayIndexOutOfBoundsException is generated. Explain what index values would be invalid and consequently would cause that exception to be generated. Be specific.

2. If more than one day has the same highest temperature, which day will the program report? For example, if both day 10 and day 20 had a highest temperature of 98, which day would the program report?

Debugging Challenge

Consider this code that is intended to print an array:

```
int [ ] array = new int [10];
for ( int i = 1; i <= array.length; i++ ) {
   System.out.print( array[i] + " " );
}
System.out.println( );
```

The code compiles without an error, but when the code is run, this error occurs:

```
Exception in thread "main" java.lang.ArrayIndexOutOfBoundsException: Index 10 out of bounds for length 10
```

What is wrong, and how can you fix it? Record your response in the space provided.

_____

_____

_____

_____

Lab 13-3

Searching Arrays

Often, the programmer wants to search for a particular item among many. If these items are stored in an array, the array can be methodically searched in order, comparing the value of each element of the array to the desired item. This process will work whether the array is sorted or unsorted. Before beginning this activity, download the files for this lab from the student companion website.

Learning Goals
- Construct a search of an unsorted array.

Materials
- jGRASP Integrated Development Environment
- Starter files from the student companion website

Application and Extension of Knowledge

Using an Array to Find Popular Baby Names

The Social Security Administration (SSA) keeps track of the names of babies as parents register a child for a Social Security number. The SSA ranks these names by popularity and posts the most popular names on its website for use by the public. In this activity, you will search an array of the most popular names for babies. The 500 most popular girl names and boy names have been stored in two text files. Each file contains names ranked in order from most popular to least popular. Each line contains one name, and all names are one word.

Your program should read the text files into two different arrays: one holding girl names and the other holding boy names. After storing the names in the two arrays, prompt the user for a name. Then, search each array for that name. If the name is found, report the popularity ranking of the name. If not found, report that the name is not in the top 500 names. Typically, a name will be in one array, but not the other, although several names appear in both lists.

The files list the names in order from most popular to least popular. Because you will read the file and store the names sequentially in an array, the ranking of a name can be determined by its position (index) in the array. The only thing to remember is the first index is 0 and the rankings start at 1. So, you will need to calculate the ranking by adding 1 to the index.

Use the sequential search algorithm presented in the textbook chapter. That algorithm can be stated as:

1. Set result index to –1.
2. Traverse the array. At each element: if the value of the array element is equal to the search key, set result index to the current index.
3. If the result index is still –1, the search key was not found. Otherwise, the result index is the index where the search key was found.

Procedure

1. Plan your solution in the space provided.

   ```
   /***** 1. Define the two arrays: one for girl names and one for boy names.
   */
   ```

   ```
   /***** 2. In a try/catch block, read each file and store the names
              in the arrays. The files are named: Girls2018.txt and Boys2018.txt.
   */
   ```

```
/***** 3. Output the most popular girl name and the most popular boy name.
          Hint: these will be at index 0.
*/
```

```
/***** 4. Prompt the user for a name to find. Then, search each array.
          Use the search algorithm presented in the chapter and summarized in
          the lab instructions. Remember that the array elements are Strings,
          so you will need to use equalsIgnoreCase method to compare the
          name to the user's request. Report results for both arrays.
*/
```

2. Launch jGRASP, and open **BabyNamesSnippet.java** starter file. You will see it includes the import statements, a constant for the size of the arrays, and a **Scanner** object for reading the name from the user.
3. Enter the code you wrote above.
4. To test the code, enter a girl name, a boy name, and a name that will not be found in either file.

Name _____ Chapter 13 Working with Arrays **197**

Reflections

1. Because array indexes start at 0 and the rankings start at 1, you need to add 1 to the index to get the ranking. You also needed to make this adjustment to the Hands-On Example 13.3.1 Movie Theater Show Times in the textbook chapter. Why do you think an array index starts at 0 instead of at 1? Is there an advantage to starting at 0?

_____

_____

_____

2. Suppose you want to view a list of the most recent names being given to babies. Visit the Social Security website (www.ssa.gov), and search for **popular baby names**. You will see a table of the top 10 baby names of the most recently completed year. The names are provided in a table with three columns: ranking (1 through 10), male names, and female names. If you download this information, how could you modify your program to use the information as it is presented rather than in two separate files?

_____

_____

_____

_____

Debugging Challenge

In your program, you write this code to define a class:

```
1  public class TrackScores {
2    private int [ ] scoresArray;   // define array
3    public TrackScores ( int size ) {
4      int [ ] scoresArray = new int[size]; // instantiate array
5    }
6
7    public int getArraySize( ) {
8      return scoresArray.length;
9    }
10   // more methods here
11 }
```

You wrote this client code to instantiate an object of the MyClass class:

```
3  public class TrackScoresClient {
4    public static void main( String [ ] args ) {
5      TrackScores ts = new TrackScores ( 10 );
6      System.out.println( ts.getArraySize( ) );
7    }
8  }
```

The class and client compile without errors, but when you execute the client code, you get this error:

```
Exception in thread "main" java.lang.NullPointerException
  at TrackScores.getArraySize(TrackScores.java:8)
  at TrackScoresClient.main(TrackScoresClient.java:4)
```

A **NullPointerException** occurs when you try to use an object reference that is **null**. The exception occurs in the **TrackScores.java** file on the line:

```
return scoresArray.length;
```

This implies the **scoresArray** reference is **null**. What is wrong, and how can you fix the code? Record your response in the space provided.

Name _____ Date _____ Class _____

CHAPTER 14

Graphical User Interface

A popular application of graphical user interfaces (GUI) is app development. Using touch or keyboard presses to make selections makes an app interactive. Input is often prompted by a Label object containing instructions. Java has a variety of input UI controls. In the textbook, the TextField class was used. In this lab, you will explore other input UI controls, animation classes, and game development.

Chapter Highlights

- In a graphical user interface (GUI), the screen displays visual options with which the user can interact.
- To promote a good feeling from the interface, user experience (UX) designers make the interface clear, easy to learn, attractive, and able to rapidly produce meaningful results.
- An event is an object created as a result of user action. An event handler is code that is executed when the event is generated.
- A button control is used to process an action when the user clicks the button.
- A label is noneditable text used to provide information to the user.
- A TextField or PasswordField control accepts input.
- A horizontal box layout will provide a side-by-side arrangement of the controls. A vertical-box layout aligns controls one on top of another.
- A radio button is a control presented in a group in which only one control can be selected. Use radio buttons when only one option in a set is allowed. Radio buttons must be added into a ToggleGroup object for the exclusive selection to work.
- A check box is a control presented in a layout group in which all, some, or none of the controls can be selected.

While studying, look for the activity icon for:
- Vocabulary terms with e-flash cards and matching activities.
- Starter files for lab activities.

These activities can be accessed at
www.g-wlearning.com/informationtechnology/1773

Warm-Up Exercises

_____ 1. Consider the following code segment. Which line(s) of code set properties of the button btnLogin?

```
22 Button btnLogin = new Button( );
23 btnLogin.setText( "Log In" );
24 btnLogin.setLayoutX( 120 );
25 btnLogin.setLayoutY( 150 );
26 btnLogin.setVisible( true );
27 effects.add( btnLogin );
```

A. All lines of code.

B. 24, 25, and 26

C. 27

D. 23, 24, 25, and 26

E. 23 and 27

_____ 2. Consider the following code segment. In addition to javafx.scene.control.*, what classes must be imported to support the following lines of code?

```
txtPlayerName = new TextField( );
txtPlayerName.setLayoutX( 40 );
txtPlayerName.setLayoutY( 50 );
txtPlayerName.setFont( Font.font( "arial", FontWeight.BOLD, 30 ) );
txtPlayerName.setTextFill( Color.ORANGE );
```

A. javafx.scene.text.* and javafx.scene.paint.*

B. javafx.scene.text.* and javafx.scene.font.*

C. javafx.scene.text.* and javafx.scene.color.*

D. javafx.scene.text.*

E. javafx.scene.textfield

_____ 3. Consider the following JavaFX code segment. What is the width of the rectangle?

```
Rectangle rect = new Rectangle( 130, 40, 58, 123 );
rect.setArcWidth( 20 );
rect.setArcHeight( 25 );
```

A. 130

B. 40

C. 58

D. 123

E. 20

_____ 4. Consider the following code segment. Which class ensures exactly one of these radio buttons is selected?

```
rb1 = new RadioButton( "New Game" );
rb2 = new RadioButton( "Saved Game" );
rb3 = new RadioButton( "Quit" );
```

A. HBox

B. VBox

C. ToggleGroup

D. CheckBox

E. Node

_____ 5. Select the method used to create an event handler for a button defined as button.
   A. button.selectEvent( )
   B. button.getHandler( )
   C. button.setOnEvent( )
   D. button.setOnRelease( )
   E. button.setOnAction( );

_____ 6. What is the visual representation of a player character in a multimedia game?
   A. ImageView
   B. class
   C. avatar
   D. instance
   E. image

_____ 7. The observable list keeps track of these items in a JavaFX scene.
   A. nodes
   B. arguments
   C. resources
   D. classes
   E. libraries

_____ 8. Which of the following methods initiates the JavaFX GUI application?
   A. start( )
   B. setScene( )
   C. launch( )
   D. effects.add( )
   E. event( )

_____ 9. Which of the following UI controls provides a side-by-side arrangement of other controls?
   A. image view
   B. horizontal box layout
   C. check box
   D. vertical box layout
   E. radio button

_____ 10. Which of the following classes is used to display an image file in a JavaFX UI application?
   A. Image
   B. Picture
   C. ImageView
   D. Photo
   E. Pixels

Lab 14-1

JavaFX Graphical User Interfaces

Applying what you learned about user interface controls, you can make apps for desktop computers and smartphones. For each destination, you would need to comply with the individual deployment processes. This is not a trivial endeavor. However, to make the apps and run them in jGRASP, all you need is JavaFX and an imagination. In this lab, you will make two apps by extending what you have learned about text, color, and shapes in Chapter 7 and extending that knowledge with the Button, Label, and additional UI controls. Before beginning this activity, download the files for this lab from the student companion website.

Learning Goals
- Evaluate a graphical user interface for effectiveness of the user experience.
- Use the JavaFX UI controls Button and Label to create apps.
- Experiment with additional UI controls.

Materials
- jGRASP Integrated Development Environment
- Starter files from the student companion website

Application and Extension of Knowledge

Designing an Age Calculator with UI Controls

In an earlier lab, you were tasked with capturing and validating user input for a date. Java has a JavaFX UI control and several classes for manipulating dates:
- DatePicker to enter a date interactively;
- LocalDate to get the date criteria for today and extract the years, months, and days from that date; and
- Period to subtract two dates in LocalDate format.

In this activity, you will write an app that uses these classes along with Labels and a Button to calculate the number of years, months, and days since someone's birth.

To avoid the struggles with data validation on date input, JavaFX provides the DatePicker UI Control. The control as it appears in the app is shown below. To enter a date, the user clicks the small calendar icon to the right of the text box. DatePicker then opens an interactive calendar. The user clicks through the years, selects the month, and clicks a day. This order is important. Then, the calendar closes, and the picked date appears in the text box in mm/dd/yyyy format.

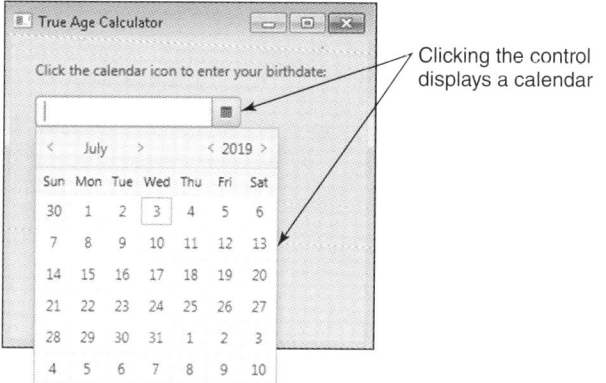

Goodheart-Willcox Publisher

Procedure

1. Apply what you learned about using the Java Class Library to find the API for the DatePicker UI control. Perform an Internet search for oracle javafx datepicker.

2. Write the API for the constructor without arguments in the space below. Identify the package that must be imported to use DatePicker. List two methods inherited from the class javafx.scene.Node that allow you to place the control at a particular location in the scene.

_____

_____

_____

_____

3. Once a date is chosen using DatePicker, the date is made available in LocalDate format. LocalDate objects are created using static methods. Find the API for LocalDate. Show the methods for getting today's date and for extracting the year, month, and day from a LocalDate.

_____

_____

_____

_____

4. Calculating the number of days between two dates is not a trivial task. Java has another time class named Period. The method between( ) calculates the number of days between two LocalDates. Locate and write below the API for finding the methods to use to extract the number of years, months, and days in the period. Make a note to import the java.time.* package.

_____

_____

_____

_____

5. Launch jGRASP, and open the TrueAgeSnippet.java file. Examine the code to see how the DatePicker, LocalDate, and Period are used.
6. For steps 6 through 13, perform the coding required and run the compiler. This will help to catch typographical errors. Locate comment #1. Label lblEnterDate has been defined. Create the object. Complete the prompt to enter a birthdate. Then set the *x* and *y* coordinates for it. Add lblEnterDate to the list of effects.
7. Locate comment 2. Add the datePicker. The variable datePicker has been defined for you. Create the object. Set the *x* and *y* coordinates for it. Add datePicker to the list of effects.
8. Locate comment 3. Define a label to display results. Label lblTrueAge has been defined for you. Create the object. Set *x* and *y* coordinates for it. Keep the label blank at this time. Add it to the list of effects.
9. Locate comment 4. Add a label to display Happy Birthday if today is the user's birthday. Nothing has been started for you.
10. Locate comment 5. Add the button to calculate True Age. The Button btnTrueAge has been defined for you. Create the object and set the coordinates. Add it to the list of effects.
11. Locate comment 6. Write a statement to set the visibility of the Birthday label to false. If the new date is not the birth date, you need to hide the message. This makes the app reusable when the button is clicked again.
12. Locate comment 7. This is where you will decide if the date provided is today. If it is the user's birthday, show your label. Use the now object from the difference in the dates calculation.
13. Locate comment 8. Ensure all your controls are added to effects.
14. Locate comment 9. Set the title for the stage to "True Age Calculator."
15. Compile and run the app. Comment on your experience below. List any errors encountered and how you handled them.

_____

_____

_____

_____

_____

Reflections

1. Reflect on the user interface of the DatePicker control. How would you improve its function?

_____

_____

_____

2. In lieu of adding a button to the app, how could you use the setOnAction( ) method for DatePicker? How would that change the app?

_____

_____

_____

Using Events to Animate Images

Similar to the notion that clicking a button generates an event that can be handled, Java provides transitions that can generate events. The idea of animation is to display a series of images with a wait time between each. In JavaFX, the PauseTransition is the class that provides a wait. In this activity, you will write a program to animate a series of images.

Procedure

1. Perform an Internet search to locate the API for the JavaFX class **PauseTransition**. Write its API in the space below. Include the setDuration method.

   _____

   _____

   _____

2. The PauseTransition class inherits from the Animation class, which generates an event when the duration of the PauseTransition ends. This is the event you will handle. It is in the event handler, defined using the setOnFinished method, that the images will be swapped. Each time the event fires, change to the next image. There are 24 images in this set of cells. Plan what you will do to keep the animation running after the 24th image is displayed.

   _____

   _____

   _____

3. Launch jGRASP, and open the BunnyHopSnippet.java file. Be sure the folder contains the 24 .png files named rabbit0.png through rabbit23.png. Compile and run the app.

4. Locate comment 1 and follow the instructions. Write your findings in the space below.

   ```
   /***** 1. Change the duration to see the effect of the timing on the animation.
    */
   ```

   _____

   _____

   _____

5. Locate comment 2. In the space below, review the steps required to add a control to an app.

   ```
   /***** 2. Add a Label UI control to display the value of currentRabbit
              below the animation.
    */
   ```

   _____

   _____

   _____

6. Add the label to the PauseTransition event handler. Compile and run the app. Write your observations below.

   _____

   _____

   _____

Reflections

1. Evaluate the quality of the animation. Does it look credible? Explain why or why not.

   _____
   _____
   _____

2. Explain how many images you think are needed to make an animation credible. Is 24 enough? Is this too few? Is this too many?

   _____
   _____
   _____

Animating Bouncing Objects

In an earlier lab, you animated a ball using a custom **Sprite** class you created. In this activity, you will bounce a box (a rectangle) off all four walls of the stage using the **PauseTransition**. The box will be located on the stage at a random location. Then, it will move both vertically and horizontally. When the box encounters a wall, the direction and color of box will change.

Procedure

1. Launch jGRASP, and open the BouncingBoxSnippet.java file. Compile and run the app. Describe the action in the space below.

   _____
   _____
   _____

2. Locate comment 1. Add horizontal movement to the event handler. Compile and run the app.

   ```
   /***** 1. Add horizontal motion. When the ball hits the wall, set the boolean changeColor
              to true.
   */
   ```

   _____
   _____

3. Locate comment 2. Change the fill color of the box on each bounce.

   ```
   /***** 2. If changeColor is true, generate a random color. Then, set changeColor to false.
   */
   ```

   _____
   _____

4. Compile and run the app. Correct any errors.

5. Suppose you wanted to move the box to a certain location on the stage by clicking that location? What would be added to the app to support that feature? Search for the phrase **javafx mouse click event** to locate the API. Apply what you have learned to add a mouse click event, and then write a handler to move the box to the coordinates of the click. Write the API for the event handling and methods you will need below. Then, add the code at comment 3.

Reflections

1. Compare the two apps for moving shapes: BallSnippet.java from Chapter 12 and BouncingBox.java from this lab.

2. Suppose you want to move the box to a random location on the stage when the box is clicked. How would the app need to be modified to support that feature?

Debugging Challenge

Consider the following code segment.

```
40 Label displayNumber = new Label( );
41 displayNumber.setLayoutX( 20 );
42 displayNumber.setLayoutY( 80 );
43 effects.add( displayNumber );
```

When the program is compiled, the following error is displayed.

```
Code segment.java:40: error: cannot find symbol
  Label displayNumber = new Label( );
  ^
```

What is wrong, and how can you fix it? Record your response in the space provided.

Lab 14-2
GUI Input

JavaFX provides many UI controls for input. Using the Java Class Library, you can learn how to implement them. In addition to the API, Oracle provides tutorials about these controls. Your foundation in this course prepares you to investigate new classes and implement them in your code. Applications for smartphones and desktops can be created using Java and JavaFX. In this lab, you will create an app to calculate the tip at a restaurant using a new UI control. Before beginning this activity, download the files for this lab from the student companion website.

Learning Goals
- Use a Text UI control and try/catch for validation.
- Obtain user input with radio buttons.
- Set properties for UI controls.

Materials
- jGRASP Integrated Development Environment
- Starter files from the student companion website

Application and Extension of Knowledge

Designing a Tip Calculator with UI Controls

At a restaurant, it is common courtesy to give servers a gratuity, or tip, for good service. A customary tip for excellent service is 20 percent. However, people may determine that another percentage is appropriate. To save time, a tip calculator can provide the amount of tip and total bill after the user enters the check amount and selects the tip percent. Your goal is to produce the app shown below.

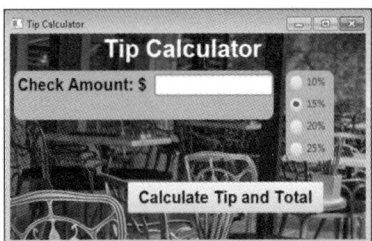

Goodheart-Willcox Publisher

Section 14.2 in the textbook introduced a large number of UI controls. You will apply what you learned to set properties and produce this app from the TipCalculatorSnippet.java starter file. There are many edits to make. Compile the app after each step to catch typos.

Procedure
1. Launch jGRASP, and open the TipCalculatorSnippet.java file. The controls are loaded, but not formatted. Become familiar with the app code. Determine the algorithm used for this app. List the steps in the space below.

_____

_____

_____

2. Locate comment 1. Notice custom methods have been created to instantiate and set properties for the controls. They are incomplete. You will complete them in the following steps. Describe the benefit of making these methods in the space below.

   ```
   /***** 1. Visit each method and complete the setup of the UI controls.
   */
   ```

   _____
   _____
   _____
   _____
   _____

3. Locate comment 2. Write code that will determine which radio button is selected and get the percentage to be used for the tip calculation. In the space below, name the Java statement you will use to make this determination. Use the double tipPercent defined at the top of the program to hold the value.

   ```
   /***** 2. Get tip percent from the selected radio button.
   */
   ```

   _____
   _____
   _____
   _____
   _____

4. Locate comment 3. The user may not input a valid value for the check amount. Perform an input validation with a try/catch block. In the catch block, output the error message. The program will continue as soon as the user presses the button again. In the space below, list the statements that will be in the try block

   ```
   /***** 3. Add try/catch to handle a NumberFormatException
           on the user input of check amount
   */
   ```

   _____
   _____
   _____
   _____
   _____
   _____
   _____
   _____
   _____

5. Locate comment 4. Apply what you learned about defining a URL resource and loading an image into an ImageView. Which method converts the resource to a String for the argument of the new Image constructor? Write the code to define the URL resource and create the new Image imgBackground.

   ```
   /***** 4. Create URL resource and load image: background.jpg.
   */
   ```

   _____

   _____

6. Locate comment 5. Assign the image to the ImageView and set layout coordinates of (0,0). In the space below, explain why it is not really necessary to set coordinates of (0,0).

   ```
   /***** 5. Set ImageView properties of image and layout.
   */
   ```

   _____

   _____

   _____

   _____

7. Locate comment 6. The Label for the title has been defined, but its properties need to be set. Set the Text to Tip Calculator. Then, set the properties for font, color, and layout coordinates. In the space below, explain why you do not need to write a setText( ) method for this task.

   ```
   /***** 6. Set properties for label: text, font, color, layout coordinates.
   */
   ```

   _____

   _____

   _____

   _____

8. Locate comment 7. The Label for displaying the result has been defined. Set its properties. Then, set the properties for font, color, and layout coordinates. In the space below, explain why you do not need to write a setText( ) method for this task.

   ```
   /***** 7. Format the Label text and set its location on the stage.
   */
   ```

   _____

   _____

   _____

   _____

   _____

   _____

   _____

9. Locate comment 8. The radio buttons for the tip percentage have been defined, but their properties need to be set. Set the Text to the percentage of tip to be calculated: 10%, 15%, 20%, or 25%. Then, set the properties for font, color, and layout coordinates. Explain why you do not need to write a setText( ) method for this task.

   ```
   /***** 8. Add text for the radio buttons to display: 10%, 15%, 20%, and 25%.
   */
   ```

   _____

   _____

10. Locate comment 9. The ToggleGroup group has been defined. Add the radio buttons to the group.

    ```
    /*****  9. Add toggle group for the radio buttons.
    */
    ```

11. Locate comment 10. Determine which of the four radio buttons should be selected when the app opens. Add the method setSelected(true) to your selection.

    ```
    /***** 10. Set the radio button that appears selected at the beginning of the app.
    */
    ```

12. Locate comment 11. Set the properties for the VBox.

    ```
    /***** 11. Set properties for the VBox layout and spacing.
    */
    ```

13. Locate comment 12. Format the button.

    ```
    /***** 12. Set properties to format the button.
    */
    ```

14. Locate comment 13. Set the properties of the error message.

    ```
    /***** 13. Set properties for the error message in case check amount is not valid.
    */
    ```

15. Compile and run. Verify all aspects of the Tip Calculator are running correctly. Enter a nonnumeric character or a $ with a number in the Amount field. What happens?

    _____

    _____

    _____

Reflections

1. Explain how you would improve this app if a diner wanted to enter a tip percentage not listed.

   _____

   _____

   _____

2. Explain how else you would improve this app. For example, should it rely on a button? Would it be better if the app updated whenever the radio button is clicked?

   _____

   _____

   _____

Debugging Challenge

This method is intended to display an error message as a result of catching an input exception.

```
public void setUpErrorMessage( ) {

   lblErr = new Label( );
   lblErr.setLabelString( "Error in Amount field!" );
   lblErr.setLayoutX( 15 );
   lblErr.setLayoutY( 75 );
   lblErr.setFont( Font.font( "arial", FontWeight.BOLD, 15 ) );
   lblErr.setTextFill( Color.RED );
   lblErr.setVisible( false );
   // add error message to the stage
   effects.add( lblErr );
}
```

When compiled, this error is generated:

```
GUIInputSnippet.java:65: error: cannot find symbol
      lblErr.setLabelString( "Error in Amount field!" );
            ^
  symbol:   method setLabelString(String)
  location: variable lblErr of type Label
```

What is wrong, and how can you fix it? Record your response in the space provided.

_____
_____
_____
_____
_____
_____
_____
_____

Lab 14-3

JavaFX Games

A key to a good game is interaction that allows players to feel in control of their destinies. You created a text-based version of the game Throw Down in a previous lab. This can be a fun and exciting game even in text-based form, but can be improved with a graphical user interface.

Learning Goals
- Reuse code from an existing game.
- Add interaction to an existing game.

Materials
- jGRASP Integrated Development Environment
- Completed files from the Laboratory Manual and Hands-On Examples from the textbook

Application and Extension of Knowledge

Adding a GUI to Create a Game App

In this activity, you will add interaction to the Throw Down game to create a graphical app. This task is a challenge. It is best accomplished with a partner or a team. Writing GUI apps uses a different thought process than making a linear Java program.

The figure below shows one possible interface for the game. You may design your own version. If you desire, add images of the dice roll by reusing code from the Race to the Top game created in the textbook.

Goodheart-Willcox Publisher

You have learned a great deal about Java and JavaFX. Now it is time for you to make a complete game from scratch by yourself or with a partner. The text-based Throw Down game you created in the Chapter 12 labs can be used as the basis for the app. You will design the interface, create the controls, assign their properties, and use a button to play in lieu of the while loop that controls the text-based version. This project is the culmination of your experience with Java and JavaFX. Have your ThrowDownSnippetPart2.java, Die.java, and Player.java files handy so you can reuse most of the code.

Procedure

1. Take time to consider the differences between the two interfaces: the keyboard app versus the GUI app. How will interaction change the experience?

2. Incorporate two **Progress Bar** UI controls in the app to visually show the progress of each player. Search the Java Class Library for the API for a Progress Bar. Enter below how you will incorporate and update the progress bar.

___

3. This game has three states:
   - get the player names and display them;
   - roll the dice and update the scores; and
   - report the win or loss.

   You can use the same button for all three states by changing the button text and using an if/else if statement to determine what to do inside the setOnAction( ) event handler method.

   Set the button text to "Play Game". After the players enter their names and click the button, if the text is "Play Game," display their names and change the button text to "Roll Dice" or "*name*'s Roll."

   If the text is "Game Over," do nothing. Do not react to a button press. This essentially ends the game action.

   Else, roll the dice, adjust the score, and update the progress bars. If a win or loss occurs, change button text to "Game Over." Otherwise, change players.

   Plan this part of the game in the space below.

___

4. Planning is essential. Design the interface on paper first. Identify the UI controls required. Then, create the controls using information from the Die and Player classes. Set the text of the button to "Player 1 Roll" or "Player 2 Roll." Use the property of the Player class to use the actual player's name in the text for the button. Create instance variables for the Progress Bars so you can adjust the widths for the correct player. Use the contents of the while loop for the button's event handler. List the JavaFX UI Controls you will use in the space below.

5. Apply all you learned in this course to make the game. Refer to earlier work and the Java Class Library. Code ideas from RaceToTheTopGame.java will be very useful. Create the game and play it with a friend to test it.

Reflections

1. Describe how you felt about beginning this project without a snippet file. Do you still feel that way after completing the game?

2. Explain how you would improve this game if you had more time.

Debugging Challenge

Consider this code segment.

```
pauseTransition.setOnFinished(
  event - >
  {
    if ( currentRabbit < ( numRabbits - 1 ) ) {
      currentRabbit++;
      imgView[ currentRabbit - 1 ].setVisible( false );
    } else {
      currentRabbit = 0;
      imgView[ numRabbits - 1 ].setVisible( false );
    }
    imgView[ currentRabbit ].setVisible( true );
    rabbitNumber.setText( "Rabbit #: " + currentRabbit );
    pauseTransition.play( );
  });
```

When executed, the following errors are reported:

```
BunnyHopAnswer.java:57: error: illegal start of expression
    event - >
          ^
BunnyHopAnswer.java:58: error: illegal start of expression
        {
        ^
BunnyHopAnswer.java:79: error: illegal start of expression
      });
       ^
3 errors
```

What is wrong, and how can you fix it? Record your response in the space provided.

Name _____ Date _____ Class _____

CHAPTER 15

Careers in Computer Programming

Computer programming is becoming a core requirement for many well-paying jobs. Coding skills are in demand across a broad range of careers in all aspects of government and business.

Chapter Highlights

- Programmers can use their strong problem-solving proficiencies to have great earning potential.
- There is a high demand for coders.
- Since programmers are needed everywhere, there is great career flexibility.
- Career plans and a results-oriented résumé are essential.

While studying, look for the activity icon 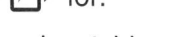 for:

- Vocabulary terms with e-flash cards and matching activities.
- Starter files for lab activities.

These activities can be accessed at
www.g-wlearning.com/informationtechnology/1773

Warm-Up Exercises

_____ 1. Which of the following is an example of a hybrid job?
   A. automotive technician using diagnostic programs
   B. estimator using a spreadsheet
   C. marketing manager forecasting inventory
   D. delivery route manager using GPS
   E. All of these.

_____ 2. Which of the following is a federal agency that measures labor market activity?
   A. Department of Labor
   B. Internal Revenue Service
   C. Bureau of Labor Statistics
   D. Bureau of the Census
   E. Bureau of Economic Analysis

_____ 3. Which programming language is recommended for database management?
   A. Java
   B. SQL
   C. JavaScript
   D. C++
   E. COBOL

_____ 4. Java shares some of its syntax with a previously developed language called _____.
   A. C++
   B. Python
   C. JavaScript
   D. BASIC
   E. SQL

_____ 5. Which of these items would *not* be included in a results-oriented résumé?
   A. salary requirements
   B. programming languages known
   C. professional accomplishments
   D. certificates
   E. formal education

_____ 6. A timeline, list of formal education opportunities, and job goals would be found in which of the following?
   A. professional accomplishments list
   B. résumé
   C. summer jobs list
   D. volunteer work list
   E. career plan

Name _____ Chapter 15 Careers in Computer Programming **219**

_____ 7. The following jobs probably belong in which career cluster: team assemblers, machinery mechanics, inspectors, and testers?
   A. Manufacturing
   B. Information Technology
   C. Law, Public Safety, Corrections, and Security
   D. Arts, A/V Technology & Communications
   E. Health Science

_____ 8. Most jobs in the Information Technology career cluster require a(n) _____ for employment.
   A. on-the-job training program
   B. high school diploma
   C. bachelor degree
   D. master degree
   E. certification

_____ 9. Which of the following websites provides up-to-date average salaries and job requirements for all jobs based on governmental data?
   A. www.monster.com
   B. www.usa.gov
   C. www.bls.gov
   D. www.indeed.com
   E. www.careerbuilder.com

_____ 10. Talking to people who can help you find a job in your field is called _____.
   A. networking
   B. strategizing
   C. discovering
   D. broadcasting
   E. working the room

Lab 15-1

Benefits of Careers in Coding

IT professionals can learn from their peers and imagine solutions they may not have thought of on their own. In this lab, you will use problem-solving skills to analyze different situations.

Learning Goals
- Identify available IT jobs.
- Express good problem-solving techniques.
- Describe positive and negative group work.

Materials
- Computer
- Internet access

Application and Extension of Knowledge

Surveying Technology Careers

Many technology jobs are available. They have a wide range of salaries and requirements. Preparing now will afford you an opportunity to set short-term and long-term goals and help you develop a career plan.

Procedure

Investigate one job-search website, such as Monster, Indeed, or USA.gov. Enter *technology* as the search term. Select three positions of varying degrees of expertise. One of these jobs will be used in Lab 15-2. Note the position's title, requirements, and salary in the following spaces.

1. Position 1
   A. Title: _____
   B. Requirements: _____
   _____
   _____
   _____
   C. Salary: _____
2. Position 2
   A. Title: _____
   B. Requirements: _____
   _____
   _____
   _____
   C. Salary: _____
3. Position 3
   A. Title: _____
   B. Requirements: _____
   _____
   _____
   _____
   C. Salary: _____

Reflections

1. Which one of these three positions would be best for someone just beginning in the IT field? Explain why.
   _____
   _____
   _____

2. Which one of these three positions has the most comprehensive requirements?

Problem-Solving with Small Groups

Programmers often work in teams. They have to be able to agree on the goal, take on a complex project, and identify the steps to solve a problem. The team members have to communicate with each other on progress as well as issues discovered. The following is a partial list of *do's* and *don'ts* when facilitating a committee or team.

Do	Don't
Call on everyone	Get too attached to your own thoughts
Allow brainstorming to find other approaches	Work on the problem alone
Use the collective wisdom of the group	Limit others' contributions
Keep track of or write any ideas on a whiteboard	Evaluate until all contributions have been made
Evaluate conditions	Settle on one condition
Compare alternatives	Give significance to different conditions
Make a decision	Worry if it is not perfect

Procedure

The following is a meeting among the programmers of Lake County Development Company to discuss which charity they want to support with their holiday donation. The figure shows how the meeting went. What are the weaknesses in this problem-solving scenario?

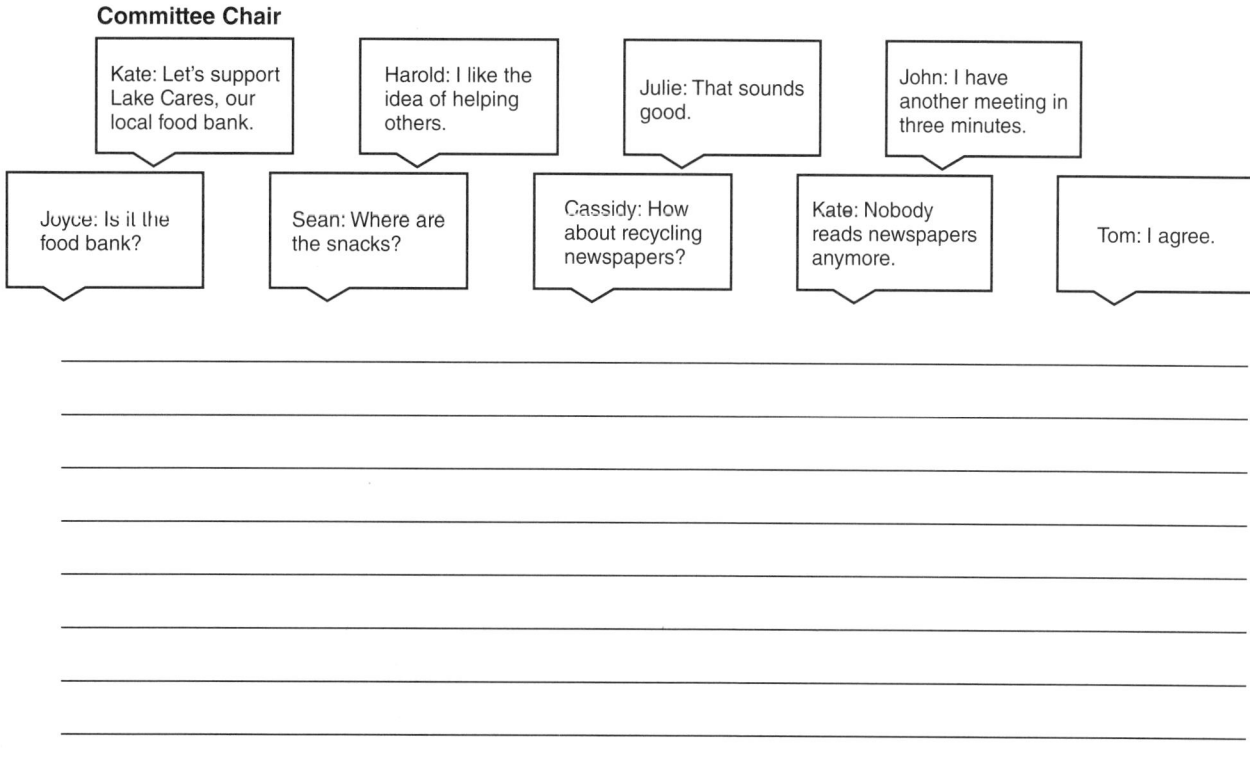

Reflections

1. Based on the table of *do's* and *don'ts*, how would you have conducted the meeting?

   _____

   _____

   _____

2. Think about a group project in which you recently participated. How did it compare to the list of *dos* and *don'ts*?

   _____

   _____

   _____

Problem-Solving with a Crowd

Groups of online communities can be used to solve problems. This is called *crowdsourcing*, a term coined by Jeff Howe from *Wired* magazine. A company or organization announces a problem online with a request for solutions. The crowd offers solutions. Then, the solutions are broadcast. The crowd develops criteria and sorts through the suggestions, eventually finding the best one. Often, there is little or no compensation for those participating in crowdsourcing.

As a company, the LEGO Group is responsible for one of the best examples of crowdsourcing. The company allows users to design new products and test the demand. Any user can submit a design for which other users are able to vote. The concept with the most votes gets moved to production, and the creator receives a 1 percent royalty on revenue.

Another example of crowdsourced problem-solving comes from the phone app called Waze. It allows users to solve problems every day by reporting traffic jams and automatically giving directions for the best route to take. Waze crowdsources information by measuring drivers' speed to determine traffic jams. It also asks users to report road closures.

The following are projects that invite a crowd to solve a problem. The first choice is a not-for-profit organization originated by a group of astronomical researchers at Berkeley. The second is a proofreading project designed to convert public-domain books into e-books. The third is a for-profit company that tests software products.

Stardust Project

Interstellar dust particles were returned to Earth by the Stardust mission. They are the first such pristine dust particles ever collected in space, and scientists are eager to investigate them. The dust particles are only about one micron (one millionth of a meter) in size. The particles were captured in an aerogel collector and set of foils. Finding them is something like searching for a handful of ants on a football field by searching one five-centimeter square area at a time. The results of the aerogel and foils will be sent to volunteers electronically to look for tracks of dust particles.

By asking for help from talented volunteers all over the world, this project can be accomplished in a lot less time than it would take a handful of scientists and lab technicians to do alone. The project website is stardustathome.ssl.berkeley.edu.

Name _____   Chapter 15 Careers in Computer Programming   223

Distributed Proofreaders

Distributed Proofreaders provides a web-based method to ease the conversion of public-domain books into e-books. By dividing the workload into individual pages, many volunteers can work on a book at the same time. This significantly speeds up the conversion process. Volunteers are given a page scanned as an image and by optical character recognition (OCR) image on a single web page. This allows the text to be easily compared to the image, proofread, and sent back to the site. The project website is www.pgdp.net.

uTest

When developers work on a product, they become highly involved in the process. They can become so immersed that they find it difficult to conduct unbiased software testing for the website or app themselves. Placing the product in the hands of a set of targeted end users before a public release offers great advantages. Issues such as crashes and confusing app flow can be noticed with the help of this type of beta testing. Feedback from potential customers without being under market scrutiny will lead to better product quality. The project website is www.utest.com.

Procedure

Choose one of the three projects described. Determine the following, and note your responses.

1. Project chosen: _____
2. What is the mission of the project?

   _____
   _____
   _____
   _____
   _____
   _____
   _____

3. What are the requirements for participation?

   _____
   _____
   _____
   _____
   _____
   _____
   _____

4. What are the potential personal or professional rewards for solving the problem?

   _____
   _____
   _____
   _____
   _____
   _____
   _____

5. Is the project a for-profit or a not-for-profit venture? How do you know this?

Reflections

1. Why did you choose this project?

2. Describe what appeals to you about the project and what concerns or reservations you would have if you were to participate.

Lab 15-2

Preparation for Careers in Coding

Preparation for a career in computer programming, and information technology in general, requires a specific set of skills. A properly prepared résumé is essential to finding a job in the IT field. In this lab, you will create a résumé that emphasizes your expertise in the IT field.

Learning Goals
- Complete the objective of your résumé.
- Display your professional accomplishments and employment history.
- Compile your references.

Materials
- Computer
- Word-processing software

Application and Extension of Knowledge

Writing a Résumé Objective

The objective on your résumé concisely tells the reader what field and position in which you are interested. It also summarizes your qualifications and personality traits that you feel are most relevant. Eventually, you will substitute "preparatory coursework" with your degrees and certifications.

Name _____   Chapter 15 Careers in Computer Programming   **225**

Procedure

1. Launch a word processor, and begin a new blank document.
2. Place your contact information on the top of the page. Include your name, physical address, phone number, and e-mail address. It can be centered or left-justified.
3. Under the contact information, left-justify the word Objective.
4. Choose one of the jobs you identified in Lab 15-1. Fill in the blanks of the following objective. Add this to the document file under the word Objective.

   Challenging position as _____ *(title of position)* where a strong interest in coding with preparatory coursework can be combined with _____ *(highlight three of your traits such as problem-solver, team player, leader, detail-oriented, perseverant, time-sensitive, planner, manager, creative, punctual, compassionate, interpersonal communicator)* can be used to provide enhanced quality to the _____ *(type of job, such as customer service, programming, technology)* of your company.

Reflections

1. Consider your personality traits. What are they? Be honest. You will try to make these traits more obvious in the second part of the résumé, Professional Accomplishments and Work History.

   _____
   _____
   _____

2. Ask your family and friends to describe you. What do they say about your personality traits?

   _____
   _____
   _____

Professional Accomplishments and Work History

The sections Professional Accomplishments and Work History are the most difficult portions of the résumé to write. It requires thinking like a potential employer. What would an employer want to see? Good attendance, punctuality, communication skills, dedication to a project or cause are some examples.

Procedure

1. In the word-processing document, begin a new line, and enter Professional Accomplishments.
2. Write at least three of your accomplishments. Accomplishments begin with an active verb such as *displayed, implemented, demonstrated, held,* or *volunteered*. After listing the accomplishment, list the result of this action. For example:

   Demonstrated a talent for coding. Result: Earned a grade of A in Introduction to Computer Science: Java Programming.

Record all your accomplishments below, and then select the three that best match your job choice. Enter those three into your document below **Professional Accomplishments**.

_____
_____
_____
_____
_____

3. Start a new line in the document, and enter **Work History**.
4. In the space below, list your employment positions as well as volunteer jobs with the most recent first. For example:

   Food Prep, Panera Bread, Parkville, MD (2020 to present)

   Volunteer, Community Clean-up Project (April 2020 and 2021)

   After you have listed your jobs, enter them into your document under **Work History**.

_____
_____
_____
_____
_____

5. If you have received any awards, start a new section in your document called **Awards**. Eventually, this section will be called **Education/Awards**.
6. In the space below, list your diplomas, degrees, and certifications as well as awards. For example:

   Eagle Scout, Boy Scouts of America, Kingsville, MD (2020)

   Dean's List, Bel Air High School, Bel Air, MD (2020)

   After listing your awards below, enter them into your document under **Awards**.

_____
_____
_____
_____
_____

7. Carefully review your résumé to check for spelling errors.
8. Save the file. Update it as you gain more accomplishments and jobs. Stay up-to-date on jobs available in your field. Check job boards. If you find a company you want to work for, keep apprised of their job board and the positions available. Then, you will be ready to apply for an IT position when one becomes available.

Reflections

1. Examine your résumé. What areas do you feel need improvement?

   _____
   _____
   _____

2. Describe how important you feel the overall appearance of the résumé is, such as the use of white space, graphics, and typefaces.

   _____
   _____
   _____

References

After you complete an interview, potential employers will probably contact your references. A list of people who know your skills and perseverance are important. These *cannot* be family members. Likely choices for references include your instructors and previous employers. If you have done volunteer work, choose the director of the project or the agency. A list of references is a separate document from the résumé.

Procedure

1. Start a new word-processing document, copy the contact information from your résumé, and paste it into the new document.
2. In the new document, enter the word **References**.
3. Think about who you would like to use as your references. Then, ask each person if he or she would be willing to be a reference for you. Often, people will be flattered that you have chosen them. However, some people prefer not to be asked to provide a reference.
4. After you have asked permission, list the names, titles, company, phone number (preferably cell), and e-mail address for each person below. Never use a physical address. For example:

Lorraine Bergkvist, Adjunct Professor
University of Baltimore
(410) 555–5555
lbergkvist@ub.edu

_____
_____
_____
_____
_____
_____
_____
_____

5. Add the references to your word-processing document under **References**.
6. Save the file.

Reflections

1. Explain why it is important *not* to use family members as references.

   _____

   _____

   _____

2. Do you feel it is appropriate to ask a reference to preview what he or she will say about you? Why or why not?

   _____

   _____

   _____

Name _____ Date _____ Class _____

CHAPTER 16
Computing and Society

When you use the Internet, you have a responsibility to the other users, and they have a responsibility to you. Users agree to abide by basic principles of ethical behavior, observe civility, and show respect for others' ideas.

Chapter Highlights

- Digital citizenship is the standard of appropriate behavior when using technology to communicate.
- Computer safety is the responsibility of students and employees.
- Computer users should protect their identities when visiting websites.

While studying, look for the activity icon for:
- Vocabulary terms with e-flash cards and matching activities.
- Starter files for lab activities.

These activities can be accessed at
www.g-wlearning.com/informationtechnology/1773

Warm-Up Exercises

_____ 1. A student downloads a copy of a movie without authorization. This is an example of _____.
   A. piracy
   B. acceptable use policy
   C. identity theft
   D. social engineering
   E. cyberbullying

_____ 2. Software that has had its code available to the public is called _____.
   A. trademark
   B. infringement
   C. open source
   D. licensed
   E. data breach

_____ 3. What is digital citizenship?
   A. Principles of right and wrong.
   B. The standard of appropriate behavior when using technology to communicate.
   C. Attempting to get private information by appearing as a harmless request.
   D. The act of making software code available to the public at no charge and with no restrictions.
   E. Using a Bluetooth-enabled card skimmer to steal data.

_____ 4. A security violation in which birthdays and social security numbers are stolen is called a(n) _____.
   A. data breach
   B. infringement
   C. cyberbullying
   D. plagiarism
   E. All of these.

_____ 5. Which of the following would be considered ethical behavior?
   A. hacking
   B. data vandalism
   C. phishing
   D. blue skimming
   E. None of these.

_____ 6. What are small text files planted by websites on a computer user's hard drives called?
   A. illegal
   B. spyware
   C. phishing
   D. malware
   E. cookies

_____ 7. A scammer called an elderly person on the phone and convinced that person to provide their birth date and credit card information. The scammer used this information to buy items from Amazon.com. What is this an example of?
   A. identity theft
   B. ransomware
   C. scareware
   D. a computer virus
   E. infringement

_____ 8. A student downloaded an attachment thinking it was part of a game. The student wound up infecting the school's computer network with a virus. What is this an example of?
   A. identity theft
   B. dumpster diving
   C. scareware
   D. Trojan horse
   E. spyware

_____ 9. A small hospital's computer network was compromised, resulting in hackers encrypting patients' files. The hospital was asked to pay $50,000 in bitcoin for the decryption of the files. What is this an example of?
   A. dumpster diving
   B. identity theft
   C. ransomware
   D. spyware
   E. antivirus software

_____ 10. Using Facebook, Instagram, and Snapchat to obtain personal details about another person is called _____.
   A. social engineering
   B. strategizing
   C. hacking
   D. broadcasting
   E. data vandalism

Lab 16-1

Computing and Ethics

Maintaining your own privacy and respecting the privacy of other users is another facet of dealing with a digitized way of life. In this lab, you will investigate maintaining an ethical approach to the Internet.

Learning Goals
- Propose ethical actions related to intellectual property and activities in cyberspace.
- Discuss current and future issues of data privacy.

Materials
- Microsoft PowerPoint or other presentation software (optional)

Application and Extension of Knowledge

Ten Commandments of Computer Ethics

Dr. Ramon C. Barquin of the Computer Ethics Institute created 10 commandments of maintaining an honest digital life. The first five are given here.

1. Thou shalt not use a computer to harm other people.
2. Thou shalt not interfere with other people's computer work.
3. Thou shalt not snoop around in other people's files.
4. Thou shalt not use a computer to steal.
5. Thou shalt not use a computer to bear false witness.

Procedure

Make up five more rules you think are ethical directives that have bearing on being a good digital citizen. List them below. If you have access to presentation software, make a slide show of all the commandments (Dr. Barquin's and yours) and share the results with your group or class.

My ethical computer rules are:

6. _____
7. _____
8. _____
9. _____
10. _____

Reflections

1. Which commandment or commandments would you have broken if you downloaded a song illegally?

2. Which commandment or commandments would you have broken if you used somebody else's research to write your paper without properly citing it?

3. Which commandment or commandments deal with copying and selling someone else's software?

4. Which commandment or commandments deal with cyberbullying?

___

___

___

Privacy

Your smartphone enables you take photographs of your friends and others easily. Social-networking sites like Facebook, Instagram, and Snapchat allow you to post those pictures online. Privacy is very hard to maintain.

Procedure

Look at the three most recent photos you have taken and posted on a social-networking site.

Reflections

If people other than yourself appear in the photo, answer the following questions. (If they are of you, go back to earlier photos until you locate five that are not of only yourself.)

1. Did you ask for consent of the other individuals in the photos before posting them?

___

___

___

2. Why or why not?

___

___

___

3. How do you think posting photos online impacts your privacy and that of others?

___

___

___

4. Describe your own privacy expectations that you hold as a result of living in a digital society.

___

___

___

5. How can you exercise good judgment when posting information and photos on social media that will become public?

___

___

___

Lab 16-2

Computing and Security

In this lab, you will analyze various examples of malware and discover what is necessary to disinfect computers.

Learning Goals
- Analyze threats to computers and users.
- Describe methods for protecting data.

Materials
- Computer with Internet access

Application and Extension of Knowledge

Protecting Your Data

Many types of threats to computer systems exist. Protection will necessitate hardware, software, and your practices and procedures. The Center for Internet Security has a mission statement to improve the overall cybersecurity of the nation's state, local, tribal, and territorial governments.

Procedure

1. Visit the website for The Center for Internet Security (www.cisecurity.org). Scroll down to the recent blog postings. Choose one to read and summarize it below.

   _____
   _____
   _____
   _____
   _____
   _____
   _____
   _____
   _____
   _____

2. While visiting The Center for Internet Security website, click on the link for MS-ISAC, the Multi-State Information Sharing & Analysis Center (www.ciseccurity.org/ms-isac/). Look at the map. What is the alert level? (See the color chart.) Is it different for different areas?

   _____
   _____
   _____
   _____

3. The following is a list of common malware. Use your Internet browser to research one of these. What does it do to your computer? How can it be removed?
   - Emotet
   - Kovter
   - ZeuS
   - NanoCore
   - Cerber
   - Gh0st
   - CoinMiner
   - Trickbot
   - WannaCry
   - Xtrat

Reflections

The following is a quote from the MS-ISAC, which tracks and investigates data breaches.

"In Q1 2019 the MS-ISAC observed a 36% decrease in the quantity of reported breaches when compared to the previous quarter and 80% decrease year to year. The education sector experienced the most breaches, accounting for 55% of the breaches in Q1."

1. To what do you attribute a decrease in the number of data breaches?

2. Why do you think the education sector experienced the most breaches?

Lab 16-3

Safe Computing

In this lab, you will analyze a privacy policy and evaluate a website for reliability of security. You will also explore attributes of strong passwords.

Learning Goals
- Analyze privacy policies.
- Evaluate websites for reliability of security.
- Formulate practices to safeguard personal information on social media.

Materials
- Computer with Internet access

Application and Extension of Knowledge

Privacy Policy

Good personal safety practices make up a large part of protecting yourself online. Make sure your social networking profiles are set to private. Sites such as Facebook and Twitter should not be allowed to display any information that can be used to identify you. Configure all your social media accounts to control your status, who can gain access to what information, who can post, and who can share account information with others.

Procedure

1. Launch a browser and navigate to Facebook.
2. Click on the Privacy link at the bottom of the page.
3. Read the information. It explains Facebook's policies on how information is received, posted, and shared; how cookies are used; and what information is seen by web surfers.
4. Do the same for LinkedIn.

Reflections

1. What steps are necessary to delete information about yourself from Facebook?

2. How does LinkedIn's policy differ from that of Facebook?

Name _____   Chapter 16 Computing and Society   237

Evaluating a Website

To develop a rubric or standard is helpful when evaluating websites. Here are some considerations:
- Does it look fake?
- Who is the author?
- Why was the site created?
- When was it published? Is it current?
- How deeply does it cover the topic?
- Does it cite its sources?

If the answer to the first question is *yes*, do not go any further. If you give one point for each of the remaining responses that was positive, you have a total of five points. A website with five points is probably a reliable website. You will investigate three websites about Velcro. You will apply the rubric above.

Procedure

1. Launch your browser and search for Velcro crop. Select the first result. Apply the rubric.
2. Search for Velcro.com. Apply the rubric.
3. Search for Velcro Wikipedia. Apply the rubric.

Reflections

1. Which site(s) scored five points? Why?

   _____

   _____

   _____

2. Which site seemed to be a fake? Why?

   _____

   _____

   _____

3. Which site would you use for research if you had to write a paper on it? Why?

   _____

   _____

   _____

Password Strength

Ultimately, the degree of protection, security, and privacy resides with the user. You need to exercise practices and techniques that enhance the security of your system. The first line of defense to protecting data safely is a strong password. There are many websites that will check the strength of a password, many for free.

Procedure

1. Launch a browser and navigate to a search engine.

Copyright Goodheart-Willcox Co., Inc. May not be reproduced or posted to a publicly accessible website.

2. Enter the search phrase Microsoft password checker.
3. Click the link for the Microsoft password checker website. It should be at the top or very near the top of the list. Notice that this is a secure web page.
4. Enter a password you use to see its strength.
5. Vary the length of the password, the characters used, and position of the characters within the password to see how the strength of the password changes.
6. Enter a password based on this passphrase: *My teacher is great!* One example is m1T3chRisGr8!

Reflections

1. How did the strength of the password vary according to length, characters, and position?

2. What is the advantage of using a passphrase?